Studying and Teaching W.C. Falkner, William Faulkner, and Digital Literacy

Studying and Teaching W.C. Falkner, William Faulkner, and Digital Literacy

Personal Democracy in Social Combination

Koichi Fujino

LEXINGTON BOOKS
Lanham • Boulder • New York • London

Published by Lexington Books
An imprint of The Rowman & Littlefield Publishing Group, Inc.
4501 Forbes Boulevard, Suite 200, Lanham, Maryland 20706
www.rowman.com

Unit A, Whitacre Mews, 26-34 Stannary Street, London SE11 4AB

British Library Cataloguing in Publication Information Available

Library of Congress Cataloging-in-Publication Data Available

ISBN 978-1-4985-4747-5 (cloth : alk. paper)
ISBN 978-1-4985-4748-2 (electronic)

∞™ The paper used in this publication meets the minimum requirements of American National Standard for Information Sciences—Permanence of Paper for Printed Library Materials, ANSI/NISO Z39.48-1992.

Printed in the United States of America

Contents

List of Figures

List of Abbreviations

Abbreviations for works by William Clark Falkner, William Faulkner, and Mikhail M. Bakhtin used in this study are listed below (the lists are alphabetized according to the assigned abbreviation). Original Dates of publication are indicated wherever citing texts are either reissued or else translated at a later date. For more bibliographic information, see Bibliography.

William Clark Falkner's Works

LBC *The Little Brick Church* (1882; Upper Saddle River, NJ: Literature House, 1970)

RRE *Rapid Ramblings in Europe* (Philadelphia: J.B. Lippincott, 1884)

SH *The Spanish Heroine: A Tale of War and Love* (Cincinnati: I. Hart, 1851)

SM *The Siege of Monterey: A Poem* (Cincinnati: By the Author, 1851)

WRM *The White Rose of Memphis* (1881; New York: Coley Taylor, 1953)

William Faulkner's Works, Letters, and Interviews

AA *Absalom, Absalom!* (1936; New York: Vintage, 1990)

FD *Flags in the Dust* (1929, originally published as *Sartoris*; New York: Library of America, 2006)

GDM *Go Down, Moses* (1942; New York: Vintage, 1990)

ID *Intruder in the Dust* (1948; New York: Vintage, 1991)

LA *Light in August* (1932; New York: Library of America, 1985)

LG *Lion in the Garden* (1968; Lincoln: University of Nebraska Press, 1980)

MOS *Mosquitoes* (1927; New York: Library of America, 2006)
SF *The Sound and the Fury* (1929; New York: Library of America, 2006)
SL *Selected Letters of William Faulkner* (New York: Random House, 1977)
WA "Wash" (1934; New York: Vintage, 1977).

Mikhail M. Bakhtin's Works

DN "Discourse in the Novel" (1975; Austin: University of Texas Press, 2011)
FNM "From Notes Made in 1970–71" (1979; Austin: University of Texas Press, 2013)
PDP *Problems of Dostoevsky's Poetics* (1963; Minneapolis: University of Minnesota Press, 1994)

Acknowledgments

Many people have contributed to making this work possible. First and foremost, I thank Kenneth Sherwood, who always gave me quite suggestive directions and appropriate guides. Without him, I could not have accomplished my goal. I am indebted to Todd Thompson and Tanya Heflin for their critical readings of my drafts and warm and acute comments. My sincere appreciation goes to David B. Downing, Gail Ivy Berlin, James M. Cahalan, Karen Dandurand, Susan I. Gatti, and Lingyan Yang as well. All of these figures above gave me proper instructions during various stages while I was developing my project at Indiana University of Pennsylvania. The faculties, staff members, and students of Seinan Gakuin University in Japan have also helped me in many ways. My colleagues in the English Department gave me significant inspirations and hints on my work. The staff members supported me as I built my pedagogical system both in class and on the net. Especially I am grateful to the students who cooperated with my project. Their positive attitudes and achievements in class firmly supported my investigation.

During my stay with my wife and daughter in the United States, numerous friends helped us with their warm companionship, including Yong and Jung Colen, Ira and Amanda Goldberg, Seungku Park and Seonmi Yoon, Ryad Ghanam and Khawla Kittaneh, Khaled and Ghada Abdel Moneim Abdel Salam, Bassam AlShraah, Wan-li Chen, Sue Cheng, Tomoko Oda and Kyle Nuske. These people's associations made my family's daily life delightful and deeply inspired my analysis on people's wise benevolence.

Finally, I want to express my love and appreciation for my wife, my daughter, my father and mother, and my brother and his family. Their hearty encouragement has supported my research activities, and their thoughtfulness has always been the origin of my hope for the future.

Introduction

Remapping Two Fa(u)lkners' Works and Literature Education in Today's Digital Environments

Time is not a fixed condition, time is in a way the sum of the combined intelligences of all men who breathe at that moment.

—William Faulkner, *Faulkner in the University*[1]

How do we study and teach the literature of the American South with ESL (English as a Second Language) students in today's digital environments? This question is necessary for current American literature teachers because Southern literary study and the digital environment have both drastically changed in the past several decades. In the introduction of *The Cambridge Companion to the Literature of the American South* (2013), Sharon Monteith suggests the contemporary efforts to remap the Southern literary landscape find that the cultural multiplicity of the American South abolishes "the traditional image of a racially bipolar South" and releases the region from the North-South sectionalism to understand it with "a changing historiography and cultural geography."[2] Freed from the nation's political dichotomy, the American South is now imagined as part of globalized space's multicultural communications. As Tara McPherson argues in *Reconstructing Dixie* (2003), today's "South" is a fiction as much as it is a fixed geographic space; people can enjoy "a simulacrum of a Mississippi Delta blues joint eating sushi and barbecue and listening to the new band of an old English pop star" in a Japanese multimedia company's party in Los Angeles.[3] In today's world of increasing globalization, time and space are mixed up enough to juxtapose a simulated Southern culture with Asian and European cultural icons. Multicultural juxtapositions take place to invite individuals to access a particular region's cultures and people. Digital space makes peoples' connections more mutual. Facebook, YouTube, and Twitter help users interactively communicate with a particular region's personal lives, performances, and utterances. In a way, these new media realize what William Faulkner (1897–1962) predicted in the passage quoted above. Faulkner supposed contemporary society was composed of multiple combinations of each individual's intellectual activity. The digital network has materialized what Faulkner imagined, making relationships more personalized and yet worldly connected. As so-called Web 2.0 development in the twenty-first century linked ESL readers to a particular region in more interactive ways, the literature of the American South is now settled on the global time and space in which every participant can communicate.[4]

Since Michael O'Brien's *The Idea of the American South, 1920–1941* (1979), scholars have explored the culture of the American South not as a solid and integrated social organism but as a product of people's imaginations to unify and comprehend its disparate realities. In the early twentieth-century's modern era, the American South was represented with the pastoral image that Allen Tate offered in *I'll Take My Stand* (1930). In Tate's argument, the region kept its "self-sufficient and self-evident" position in America.[5] This local image was criticized but repeated in W. J. Cash's *The Mind of the South* (1941). In his influential work the American South was expressed as "a self-contained and largely self-sufficient little world."[6] What today's scholars want to reconfigure is this traditional image. The American South is not the self-sufficient, monolithic community with an innate identity, but the place where various voices compete to fabricate its local specification in a conscious effort to promote the region to the world. In his landmark work, *Inventing Southern Literature* (1998), Michael Kreyling declares that Southern identity is a cultural product some chosen intellectuals have produced in the same way as "all historically indigenous social elites produce ideological realities."[7] Social elites create the American South's identity in order to retain their power in cultural, political, sexual, and economic areas, and then they reproduce its image as a natural one.

The literature of the American South is not the place where the monolithic white community holds its inherent privilege, but is instead the place where intellectual efforts are given to establish an imagined community in rivalry with other worldviews. Other viewpoints against the dominant discourse have frequently been associated with African Americans. In *Dirt and Desire* (2000), Patricia Yaeger argues, "southern literature, at its best, is not about community but about moments of crisis and acts of contestation, about the intersection of black and white cultures as they influence one another and collide."[8] The American South is a place where people are always aware of their socially stratified structures; people are consciously controlling their positions to deal with other social groups. Richard Gray alleges that the American South is a place of "self-fashioning." He says Southerners are full of "a matter of *language* and *communal ritual*, the human habit of positioning the self with the help of the word and others."[9] People are socially separated, wearing masks or creating temporal agreements. Sometimes the racial reality of the American South betrays the fact that the romantic unity between the hero and the heroine is an illusion. In *Light in August*, Joe Christmas is called "Negro! Negro! Negro!" by the white woman, Joanna Burden, in their bed (LA 590). This scene implies not only that racial segregation enters into lovers' private beds, but also that people cannot escape their social awareness even in their intimate relationships with sexual partners. Now the American South's multiple cultures and racial conditions function as symbolic tools or typical samples for international readers to investigate how humans exist as social beings in every aspect of life.

Recent scholars have tried to understand the American South's communities and individuals as people's efforts to position themselves among others. ESL students also take part in this by producing their connections of individuals and communities in this imagined space. In the same way Shreve McCannon, a Canadian student who tried to understand the Southerners in *Absalom, Absalom!*, can take part in the dialogue about the South's past, every ESL student can challenge Quentin Compson's claim—"You cant understand it [the South]. You would have to be born there." (AA 289)—in order to have a dialogical relationship with this uniquely punctuated area. As Kreyling suggests, in order to reconfigure cultural identity as "imaginary," others' viewpoints are effective in finding a particular region's culture as "an arbitrary set of social formations" that have been produced to be shared both with the people born there and others as well.[10] Foreigners' explorations of a particular region resist the idea that cultural identity is an innate phenomenon. Traditionally, as Cash argued, it was believed that "the dominant trait" of the Southern mind was "an intense individualism."[11] However, today the American South appears as a place where people focus on their private connections to create their own space, and their communal relationships already contain multicultural and intermixed elements that have been brought from around the world.

This study uses a coined term, social combination, to reconsider ways to study, contextualize, and teach the literature of the American South in the internationalized university's digital conditions. Social combination means two or three people's temporal connections to redeem and reinvigorate the egalitarian spirit that has been threatened by modern society's imperial impulse. Social combination is defined by three features (see figure 0.1).

These three features of social combination will be applied any time I mention the term. Social combination creates a dialogic space in which the participants see themselves in relationships with the others. In this dialogic space, one realizes one's self as not "I" but "me." In my argument, "me" is treated as an important utterance, because it suggests a character sees one's self from the interpersonal viewpoint that is created by one's communication with the other.[12] Characters' democratic, equal relationships are suggested in their autonomous actions to do something for someone else. The concept of social combination is applied to only two or three individuals' relationship

What Is Social Combination?

Social combination is considered as a model of personal relationship in today's globalized world. This personal relationship is defined by three features:

1) Social combination is an interpersonal communication among two or three people.

2) In social combination, the personal communication continues for a certain period of time.

3) Social combination aims to attain an equal relationship among two or three people for as long as their relationship continues.

Figure 0.1 Three features of social combination.

to focus on this personal sense of equality in their particular social condition. Sometimes this personal sense of equality arises in quite a short time and immediately reveals that participants are not completely free from their society's prejudices and disparities, but it has an enduring effect on every participant's social life. In this democratic relationship, characters discover the alternative functions of their selves in their relationships with the others, even if their relationships do not last long.

This study uses social combination as a beneficial concept to explore and teach the literature of the American South to ESL students. This term explains an imagined personal relationship that some literary characters find as their social identity. In the literature of the American South, the characters' social combinations are used as ideological foundations in order to claim they are living in the less imperialistic, less capitalistic, less nationalistic, perhaps less racist culture.[13] Southerners find their appeals of personal relationships are useful to protect their particular experiences from the infestation of the outer world's unsavory noises. Especially, the social combinations of African Americans and the socially marginalized whites symbolize equal relationships' temporal appearance. Typical examples are the relationships between Jim and Huck in Mark Twain's *Adventures of Huckleberry Finn* (1885) in the nineteenth century and Lucas and Chick in Faulkner's *Intruder in the Dust* (1948) in the twentieth century.[14] Other various personal relationships of socially marginalized members in the literature of the American South from the middle of the nineteenth century to the middle of the twentieth century include: brave heroine Capitola and reformed rascal Black Donald in E.D.E.N. Southworth's *The Hidden Hand* (1868–69); old slave Uncle Remus and his master's seven-year-old boy in Joel Chandler Harris's *Uncle Remus: His Songs and His Sayings* (1880); bourgeois wife Edna and independent pianist Mademoiselle Reisz in Kate Chopin's *The Awakening* (1899); private tutor Miss Minerva and her small student William in Frances Boyd Calhoun's *Miss Minerva and William Green Hill* (1909); a Southern white narrator and

African American carriage driver Uncle Caesar in O. Henry's "A Municipal Report" (1909); Virginia woman Dorinda Oakley and her club-hoot male mate John Abner in Ellen Glasgow's *Barren Ground* (1925); middle-aged African American Janie Crawford and her best friend Pheoby Watson in Zora Neale Hurston's *Their Eyes Were Watching God* (1937); and ambitious politician Willie Talos and his right hand Jack Burden in Robert Penn Warren's *All the King's Men* (1946). In these works, social combination is described as the American South's unyielding personal value that countervails the region's poverty, racism, patriarchy, white supremacy, and imperialistic arrogance.

From the second half of the nineteenth century to the first half of the twentieth century, the literature of the American South promotes the imagined personal connection as a key notion to save its culture from its social, political, and economical defects. This sanctifying of personal relationship is, however, sometimes used for white supremacists' ideological purposes. Some romantic novelists in the American South use the characters' sentimental relationships to promote the South's imperialistic dominance of Anglo-Saxons and to legitimize Southern plantation culture as the inheritor of the United States' true origin. Those characters' personal relationships cannot be called social combinations but should be called romantic unities. These novels' ideological hierarchies between characters are symbolized by certain patterns: the idealized unity between the white male master and his faithful black servant, white males' homogeneous brotherhoods, and white gentlemen's marriages with ever dutiful white ladies. The relationship between the wealthy planter Moreland and his loyal slave Albert in Caroline Lee Hentz's *The Planter's Northern Bride* (1854) is a typical romantic unity between the master and slave in the Southern antebellum era. In Thomas Nelson Page's "Marse Chan: A Tale of Old Virginia" (1887), the old former slave's memories are told in order to confirm that his young master's life is identified with his life's best period. Thomas Dixon Jr.'s bestseller, *The Clansman* (1905)—which was adapted in D.W. Griffith's film, *The Birth of a Nation* (1915)—illuminates the intimate connections of white elites: the Northerner Phil and the Southerner Ben and their marriages to two obedient sisters are a core of the unity of the white community. Their restoration of imperial dominance of African Americans in the South celebrates the rebirth of Anglo-Saxon imperial dominance in the nation. Ideologically, in Hentz's sentimental novel, the unity of the white master and his loyal servant is admired as a better relationship than the capitalists' cruel exploitations of the poor laborers; in the novels of Page and Dixon, the struggle of the South's reconstruction era is propagated as the historical pivot for the preservation of American white supremacy.

The later nineteenth century's romantic novels of Faulkner's great-grandfather, William Clark Falkner (1825–1889)—henceforth called W.C. Falkner—belong to these romantic works that insist on the value of Southern tradition and white supremacy. In his last novel, *The Little Brick Church* (1882), W.C. Falkner depicts a plantation family that uses African Americans as slaves in the Hudson River area during the colonial period, as if the origin of the Southern plantation culture could be found in Colonial America's prototypical society. This work narrates America as the country that legitimizes the Anglo-Saxon patriarch's will to protect and dominate the other socially marginalized members of its narrow community. The novel's romance between the Northern planation family's son, Oscar, and the daughter of an English aristocrat, Olivia, illustrates that America's revolutionary spirit is nurtured in a wealthy plantation family and its culture, in which the hero exercises his brave spirit to overshadow England's feudalism and to protect the weak and the helpless from the old world's evil design. W.C. Falkner's rhetoric intentionally situates the plantation culture and its romantic atmosphere as one of the origins of democracy in the United States. The characters' personal relationships are used as an ideological unity to promote the American South as an inheritor of the original social form of the nation.

This study examines the literary works of both Faulkner and his great-grandfather, W.C. Falkner, to explore how these two Southern novelists use characters' personal connections to promote their region to readers both inside and outside of the American South. I compare several of their works in the following argument, but conclusive contrasts can be found in W.C. Falkner's *The Little Brick Church* and Faulkner's *Intruder in the Dust*. Both published these novels as successful, established authors—in W.C. Falkner's case, *The Little Brick Church* was published following his bestseller novel, *The White Rose of Memphis* (1881), and in Faulkner's case, *Intruder in the Dust* was his first popularly accepted novel after he established a reputation among critics by publishing *The Portable Faulkner* (1946). After their discovery of a broader audience, both Fa(u)lkners used their characters' personal connections as an appeal of their worldviews, although their purposes are almost opposite. In W.C. Falkner's case, the romantic unity of a narrow community dominated by white male adults in the American South is promoted as the ideal goal Americans have tried to realize from its start. In contrast, in Faulkner's historical imagination, characters start by recognizing that their actual world prevents them from realizing such a romantic goal. There are some fatal faults in their innate personalities, but the characters struggle to create social combinations in order to make their mutual conditions better. Faulkner uses the socially marginalized characters' personal combinations to promote Southern culture not as the monolithic legacy of plantation culture, but as an incessant process by which individuals form social connections to resist the white male adults' imperial dominance.

I treat Faulkner's work as a historical product that uses his great-grandfather's discursive romantic image as the characters' authoritative core.

W.C. Falkner's work shows a typical white, male-oriented ideology that is similar to his contemporary popular novelists, Page and Dixon. They believe that the human beings are naturally unequal, and the South honestly follows this imperial desire to create its culture. In their literary works, they insist that the South inherits this American hidden desire. What they want to do is to follow the instincts of American imperialism, which has never been asserted in the Declaration of Independence and the Constitution. To depict the unity of the hero and the heroine in a narrow community as if they were the king and the queen in a nation is the ultimate goal of W.C. Falkner's romantic story. This pattern makes W.C. Falkner's work a part of typical imperial romance that broadly shared Euro-American culture from the late nineteenth century to the beginning of the twentieth century. This romance approves of the imperialistic hero's arrogant dominance of his occupied area. Faulkner uses this cultural legacy as a presupposed heroic image of Southern white males. This means that, in spite of the complex structure of his work, the psychological core of Faulkner's white male characters is quite typical and easy to understand for the world readers who experienced Euro-American imperialism's colonizing/colonized condition. In his early and middle works, Faulkner's white male protagonists fail to create their social combinations because this imperial-romantic pattern is still lurking at the bottom of their self-image. In contrast, in his later works, after *Intruder in the Dust*, Faulkner depicts the socially marginalized people's possibility to have equal relationships in their personal space in order to resist the white males' romantic imagination. By setting W.C. Falkner's romantic pattern in his own works' core, Faulkner dramatizes his characters' dialogical space to resist this romantic pattern's centripetal gravity.

Faulkner scholars have studied these two writers with a curious inconsistency. Many Faulkner studies find W.C. Falkner's biographical episodes have a large influence on Faulkner's fiction but consider his literary works almost worthless, as if his fiction left no effect on his great-grandson's works. W.C. Falkner's life is treated as the major origin of the world-famed author's imagination, while his romantic work in the nineteenth century is quite underrated compared to Faulkner's modern literary achievement. Donald Philip Duclos's *Son of Sorrow* (1961) is the only full-length study of W.C. Falkner's biography and his works. Duclos explores W.C. Falkner's works with high evaluation, but his study is not theoretically developed enough to connect W.C. Falkner's literary works to his great-grandson's fiction.

Two recent scholars' studies, Joseph R. Millichap's *Dixie Limited* (2002) and Taylor Hagood's *Faulkner's Imperialism* (2008), examine the relationship between both Fa(u)lkners' fiction to give a more proper perspective on their works. These scholars suggest that readers have overlooked the South's romantic novels because of the successful promotion of the twentieth century's Southern Renaissance. As Millichap says, "Contemporary revisionists of Southern literary history, such as Michael Kreyling and Patricia Yaeger, complain correctly that the Southern nineteenth century is obscured and/or undervalued in order to draw a sharper contrast with the putative brilliance of the Southern Renaissance." Millichap insists the Southern Renaissance is the twentieth century's most successful literary movement that deserves readers' special attention, including cultural heroes such as Allen Tate, Faulkner, and Robert Penn Warren. It overshadows the previous generations' authors in the American South. A way to reform this strongly biased attitude is to give proper attention to the neglected Southern authors' works. As Millichap argues, W.C. Falkner "began his own literary transformation into a Southern archetype" before Faulkner was born. Supposing W.C. Falkner's works as Faulkner's "metaphorical inheritance," Millichap reevaluates W.C. Falkner's fiction as "a sort of Rorschach blot in which the younger writer might discern the cultural baggage of his patrimony—both the creative urges and dark sins of the fathers, with their inevitable heritage of guilt, confusion, and conflict."[15] Millichap's book does not further explore the connection between W.C. Falkner's literary work and Faulkner's writing, but it does examine the railroad W.C. Falkner built as a material infrastructure that defines the course of Faulkner's imagination.

In *Faulkner's Imperialism*, Hagood studies Faulkner's *Mosquitoes* (1927) and W.C. Falkner's *The White Rose of Memphis* (1881) as works that inherit the literary tradition of Boccaccio's *Decameron* and Chaucer's *Canterbury Tales*. In both Fa(u)lkners' works, the Southern people, who escape their regular lives, find storytelling and chatting as the main pastime on the ship. W.C. Falkner focuses on the romantic narratives during a Southern white masquerade party on a luxurious ship called the *White Rose of Memphis*, which is sailing down the Mississippi from Memphis to New Orleans; Faulkner uses social satire to describe the decadent artists' gathering on the boat called *Narcissa*, a luxury yacht that is stranded off the shore of Lake Pontchartrain in New Orleans. Hagood argues that the fear of the yellow fever works as an implied historical context in these works. In the latter half of the nineteenth century, the yellow fever presents a real threat to travelers in New Orleans. According to Hagood, W.C. Falkner focuses on wealthy Southern whites as "aristocratic models" who can silently ignore the fear of epidemic: "What the reader does not get from [W.C.] Falkner are narratives that detail the plight of blacks or poor whites unable to escape the city and therefore susceptible to yellow fever." In *Mosquitoes*, Faulkner also describes the fear of yellow fever at a metaphorical level. The mosquitoes—which are the title of the novel—carry the yellow fever in New Orleans. Hagood argues, "Faulkner seems to have realized the implication of racial as well as sexual mixing in the mosquito's occupation: it brings a disease that makes one yellow, the color of mulattos, quadroons, octoroons, the color of hybridity." In Hagood's interpretation, Faulkner designs New Orleans as "a space of racial and cultural amalgamation" in order to dramatize the constant struggle between the subjugators and the subjugated. Assuming the struggle between the oppressor and the oppressed is the essential element in these fictions, Hagood states the historical context shared in both works is the important element we can use to analyze their differences. By focusing on the common historical backgrounds, Hagood argues readers can find that both authors' imaginations develop in opposite ways in order to deal with external threats.[16]

Millichap and Hagood both discuss the historical contexts that connect W.C. Falkner's works with Faulkner's fiction. Millichap focuses on W.C. Faulkner's material legacy, the railroad, as a foundation of Faulkner's imagination. Hagood argues that both author's contexts are a historical condition of the yellow fever, though it is not directly mentioned in their literary works. Both studies focus on the indirect effects of historical facts. However, neither study argues the thematic connections of Fa(u)lkners' works with a coherent perspective. The main themes of both Fa(u)lkners still need to be explored in order to understand how these authors develop their imaginations to disseminate the values of regional characters' personal relationships.

In this study, I use W.C. Falkner's *The Little Brick Church* and Faulkner's *Intruder in the Dust* as examples of the authors' ideas on democracy. Although both W.C. Falkner's *The Little Brick Church* and Faulkner's *Intruder in the Dust* dramatize their characters' personal demands for democratic relations as their main plots, their conclusions make a strong contrast. In *The Little Brick Church*, the hero and the heroine's ideal condition is crystalized in the hero's complete dominance of the heroine and his will to protect the other socially marginalized people against imperial dominance. In contrast, the socially marginalized characters in *Intruder in the Dust* discover their ability to have social combinations in the middle of their struggle against the white, male-oriented society's imperial impulse. Scholars have not considered these works as either Fa(u)lkners' best works,[17] but this study regards them as good introductions for ESL students to explore how Southern authors strove for democratic communality in their writing. I argue the social combinations in Faulkner's works are the social form that resists the romantic unity that is represented in W.C. Falkner's ideal characters.

The concept of social combination is taken from the characters' personal connections in Faulkner's *Intruder in the Dust*, and is also supported by Mikhail M. Bakhtin's concept on dialogue in "Discourse in the Novel." Bakhtin's notion of people's linguistic interactions starts with his recognition that language cannot be a unitary whole. A person's utterance in one context is different from the same person's utterance in another. Just as utterances are always the results of some social contexts, people's words come from various social backgrounds to form "the co-existence of socio-ideological contradictions" (DN 291). Language keeps this coexistence of mutually contradictive worldviews in its heteroglossia—which refers to people's irreducible diversity.[18] All heteroglot words "intersect with each other in many different ways" (DN 291). Creative novelists juxtapose these different voices to create "intentional diversity of speech" and to attain "a unique artistic system, which orchestrates the intentional theme of the author" (DN 294, 299). In modern novels that criticize romantic novels' patterns, characters' words and utterances are juxtaposed to produce the temporal resonances between characters that originally belonged to totally different social strains.

The three features of social combination—interpersonal relations, temporal continuations, and equal conditions—correspond with Bakhtin's idea of "a multiplicity of social voices and a wide variety of their links and interrelationships" (DN 263). Bakhtin finds the dialogic space in which multiple voices relate to each other in the modern novel, which presupposes that common ground among individuals will continue for a period of time. Indeed, the modern novel has established its function in order to explore common ground in every relationship since its readers have historically appeared with democratic prospects on their reading. As J. Hillis Miller argues in *On Literature* (2002), Euro-American literature, which is associated with the gradual rise of literacy in Western culture, has been connected with the growing appearance of democracies. Modern democracies have developed alongside universal literacy education, and this modernization has also allowed citizens access to printed materials. Miller argues this development of free access to texts that is connected with the world's democratic revolution continues with today's expansion of the Internet and digital technology:

> No technology has ever been more effective than the printing press in breaking down class hierarchies of power. The printing press made democratic revolutions like the French Revolution or the American Revolution possible. The Internet is performing a similar function today. The printing and circulation of clandestine newspapers, manifestoes, and emancipatory literary works was essential to those earlier revolutions, just as email, the Internet, the cell phone, and the "hand-held" will be essential to whatever revolutions we may have from now on.[19]

Historically, modern readers' identities are based on their democratic attitudes toward the world's texts. Readers' demands for free access to information propel the development of information technologies that convert printed papers to digital texts. This demand for democracy through media has never been completed even in today's digital age, and we still read modern literature for the sake of the temporal realization of freedom based on the democratic condition in the world. The readers of Euro-American literature take part in this democratic process when they read both printed and digital texts. For this modern egalitarian self, once a regional or personal matter is written and circulated, it means it is recognized as a part of the process of our world's unfinished equalization. The world always creates new stratifications, transgressions, and antagonisms, and we always need new texts to talk about them and to deal with others equally.

This idea strongly contrasts with Hegel's idea of a human mind that goes through a dialectic progress to gain unity with the universal spirit. Bakhtin argues that the whole world is something bigger than an individual's idea, believing that Hegel's idea ultimately denies the existence of others. Bakhtin's dialogism, which presupposes socially stratified voices have multiple links and interrelationships, is an antithesis of Hegel's desire to totalize the world with one unified voice. Bakhtin criticizes Hegel's dialectic development of the mind as monologism, which is defined as an extreme form of narrative in which the narrator is essentially identified with the author and says the author's own self-affirming words. In *Mikhail Bakhtin: The Dialogical Principle*, Tzvetan Todorov writes Bakhtin is "quite suspicious of Hegelian dialectics, in any case, taking exception to its desire of unifying everything—he refers to it as 'Hegel's monological dialectic,' and he speaks of the 'monologism of Hegel's *Phenomenology of Spirit*.'"[20] Bakhtin's dialogism, which presupposes socially stratified voices' multiple links and interrelationships, is an antithesis of Hegel's desire to totalize the world with one unified voice. According to Bakhtin, since words contain other speakers' perspectives, a person constantly shares ownership of his own voice with others:

> The word in language is half someone else's. It becomes "one's own" only when the speaker populates it with his own intention, his own accent, when he appropriates the word, adapting it to his own semantic and expressive intention. Prior to this moment of appropriation, the word does not exist in a neutral and impersonal language (it is not, after all, out of a dictionary that the speaker gets his words!), but rather it exists in other people's mouths, in other people's contexts, serving other people's intentions: it is from there that one must take the word, and make it one's own. (DN 293–94)

In a dialogic relationship, one's words interact with the other person's worldview to create its meaning. Bakhtin's dialogism supposes language is a common medium that resonates with people's social coexistences. In a modern novel, characters' words do not only reflect the speaker's social position, but also interact with the listener's viewpoint. The novelist is an artist who can juxtapose these dialogic relations "for the orchestration of his themes" (DN

292). The novelist's intention creates a "common plane" (DN 293) on which various voices can be juxtaposed.

While Bakhtin's goal is the uniform dialogism of the entire world, I limit the potential of my idea, presupposing that what we can attain as our common ground is a temporal, equal communication of a few persons. This limitation is added because of the current political condition—our democratic condition that should first be realized in our daily life. Our everyday democracy shows a strong contrast with Bakhtin's one. Bakhtin imagines a democratic condition as the ideal one that should be imagined in fiction. As Ken Hirschkop argues in *Mikhail Bakhtin: An Aesthetic for Democracy* (1999), Bakhtin tried to imagine a democratic culture without democratic politics because Russia did not have democracy "as a set of observed political procedures."[21] In contrast, today's people who are digitally connected to the world begin their realization of democracy in their everyday actions. Instead of Bakhtin's idealistic argument on a democratic culture, I argue that our globalized condition cannot imagine a common plane for everyone, even in a novel. The globalized world is a deeply stratified one, and people's equal relationships are quite changeable according to our specified situations. As Amartya Sen suggests in *Identity and Violence* (2006), sometimes we experience drastic shifts in our own selves or feel the existence of our "plural identities"[22] in order to have proper connections with others that have different backgrounds in our broadly connected world.

The number of dialogical participants in social combination is restricted to two or three, because this term is used to focus on this personal dialogue's social function in the modern novel. In my concept of social combination, a two- or three-person group's dialogue does not mean to say that they have conversation in an isolated, noiseless environment, but rather insist that their conversations reflect the world's various voices. In the history of the literary study, this personal dialogue's social connection with its current environment is one of the important themes. For example, in *The Dialogical Self* (1993), Hermans and Kempen describe Joseph Frank's concept, "spatial form," as "one of the cornerstones for a theory of modern novelistic literature," assuming that Frank's concept of the individuals' personal conversation in the contemporaneous society is "intimately related to Bakhtin's thesis." Hermans and Kempen explain that "Frank took Flaubert's famous country fair scene in *Madame Bovary* as a starting point in his analysis," and summarize that, in Frank's argument, Flaubert sets this important scene's action to be going on simultaneously on three levels: the surging and jostling mob in the street on the lowest plane; speechmaking officials in front of the attentive crowd on the slightly uplifted platform; Rodolpho and Emma watching the fair and carrying on their amorous conversation on the highest level. The lovers' seductive conversation ironically proceeds in the middle of various voices of the mobs and speakers.[23] As Flaubert later wrote, "*Everything should sound simultaneously.*"[24] To express the whole meaning of this situation, a dialoguer's utterances are not only the reflections of the partner's, but also contain the disparate resounds of the surrounding people's voices. This argument suggests that a two- or three-person group's dialogue in the modern novel is not just a purely private one, but an intentionally circumscribed condition that is particularly mentioned to crystalize the process of the intersubjective creation of words, that richly contains the respirations of the ongoing world.

Today, the informative condition has become more networked, and people's personal dialogues reflect the world's voices and globally expanded network's stimulations. Because of this intermingled condition, a two- or three-person group's dialogue has the potential to create some discursive words that reflect the ongoing world's condition. The various aspects of our selves are the result of multiple dialogic relationships with others. As social combinations are personal and temporal experiences with multiple variations, it is literature's task to reproduce or reimagine these relationships to increase people's ability to comprehend their value. The literature of the American South offers rich texts from which ESL students may learn how people conduct social deals to produce their own words and to evaluate their own personal equalities in their multiple connections.

AN AMERICAN LITERARY TEACHER'S EXPERIENCE WITH 9/11: AN EPISODE

Social combination is a term used to evaluate people's egalitarian efforts in their miscellaneous relationships in today's globalized world. This term also evaluates people's temporal relationships that have not attained equality despite their efforts—the budding actions that can be estimated as efforts to find common ground between socially stratified groups' members. This evaluation focuses on the discursive traces of people's sporadic actions to create common ground between individuals. As Foucault suggests in *The Archaeology of Knowledge*, focusing on the discontinuous, temporal formation is effective to decipher the historical meaning of a certain person's "particular discursive practice."[25] My argument specifies individual efforts for equal relationships as people's temporal ventures that take part in the discursive formations of a global, democratic society. Social combination starts with the notion that equality for all people is not an obvious goal because people always diversify their processes to realize their equality.

This study assumes that having equal relationship among a few individuals for a while is social combination's final goal, to cope with today's pessimistic recognition that a unified common ground throughout the world cannot be realized after September 11, 2001. After the tragic day, equality between individuals becomes a quite personal matter, and it is important to focus on what kind of equality people want to create in their temporal relationships. My experience during that tragic day and the aftermath may give a historical perspective to explain why an international American literature teacher demanded a new term, social combination, to evaluate people's temporal egalitarian activities. I was in Santa Cruz, California, teaching/facilitating a short English language program at UCSC (the University of California, Santa Cruz) for my women's junior college in Japan, when the attacks on the World Trade Center and the Pentagon occurred. My wife's very early morning call from Japan made me switch on the TV. Clouds of smoke billowing out of one of the twin buildings appeared on the monitor. I watched it for a while, hearing detailed pieces of information from my wife on my cell phone, sometimes noticing unusually emotional exclamations of American announcers on TV. I watched another plane hit the second tower. I continued the conversation with my wife. One of the buildings collapsed. I heard the broadcaster's embarrassed exclamations repeated—"Incredible, incredible!" I had a slightly strange feeling when I realized that I was hearing the more comprehensive, objective information from Japan. My wife relayed to me what she was hearing from Japanese broadcast news while I received emotionally paralyzed words and shocked what-is-now-going-on exclamations on TV.

As is often the case with such disasters, my family was divided. At the end of our long conversation, my wife asked me when I would return to Japan. I didn't know. Airports were the most dangerous places to visit. I went out of my room to consult with our program's Japanese agent who stayed in the same building before breakfast, and decided to stay in Santa Cruz for a while. "Don't worry. Santa Cruz is in a suburban area," the agent told me. "Maybe we are safer here than in San Francisco." Having no effective way to cope with my vague anxieties for the future, I had no choice but to continue my routine. I set off for the language extension center earlier than usual that morning. The fifteen-minute walk from my accommodation facility to the language center was encompassed in the fog of California's late summer. It was chilly. On my way there, I found several young people gathered in front of the Santa Cruz community paper newsstand as if they were waiting for something. I could see their vulnerable countenances through the mist. There were no iPhones, no Facebook, no Twitter at that time. Though the Internet had developed enough for most communication, in such a case of emergency many people found that the Internet—Web 1.0—could not give them what they wanted. Everyone wanted an emotional ease, a personal interaction with the world, and a sense of belonging to the community. But Web 1.0 and mass media only increased the divide between the watched and the watching. The local people still depended on the community paper for answers, even if they knew it could not ease their fears.

When I arrived at the extension center, I found almost all the staff and students were there. In a low voice, I spoke with them about whether or not we should change our program, but no alternative plans emerged to handle our anxiety. We decided to continue the normal program, and I was expected to explain what had happened to the Japanese students in Japanese and to be with them as much as possible during their classes. The students normally came to the center from their host families' houses. Almost all of them had already witnessed the terror attacks on TV, but could not understand what was going on. Of course not. No one could fully comprehend it. I just repeated the information I had heard from my wife that morning. I sat with the students while they took their ESL lessons. While I observed my students' class, I questioned myself—how could I keep my ordinary life after this incident? In what way could I imagine the world? I didn't have an answer. Up till that day, we optimistically believed the world would continue and someday we might share common ground when communicating with each other, in the same way the people demolished the Berlin Wall in 1989. Now I recognized that I woke up this morning to find our vague hope was entirely crushed. The world was just a gathering of various shelters. I was luckily in one of the shelters that was still intact. I wanted some lucky charm to protect my fragile shelter. I wanted some ideology, some religion, some sense of belonging—something to protect me from this profound uneasiness.

Many Americans soon found their charm to symbolize their belonging. Within a few days, US flags appeared everywhere, demonstrating the Americans' will to resist terror attacks. I found the flags on Santa Cruz souvenir shops, the hoods of cars, hats, T-shirts, and TV programs. The flag is the symbol of American democracy; it declares that the wearers belong to the United States, and they would live and die for the American spirit. I envied them. I wanted such a charm. I wanted to stick that flag on the top of my head as if I was an American citizen. It was one of the strangest sentiments I had ever had. As I realized the potential reality of my death in America, I wanted to be an American citizen to give my death a rightful purpose. "But I am not an American citizen," I thought to myself, "I cannot have the American flag, American democracy, and the American nation as mine. I am just a foreigner. I am one of the others in this nation." Maybe I was strongly obsessed with my responsibility. I had to take my students back to Japan in a few weeks, but nobody knew which airport and which airplanes would promise us a safe trip. No explanation was given for this international issue's cause and effect. This was not my destiny. This was just an accident. I happened to come to Santa Cruz and witness the terror attacks on the TV, and without comprehending the whole situation, I was just afraid of my and my students' meaningless death in America.

I was in America, but I found I could not share the same sense of belonging with the American citizens during this disaster. In *Imagined Community*, Benedict Anderson argues a nation is an imagined community that is shared by people who have common recognition on their belonging. If Anderson's argument is applicable to our globalized condition, a person who stays in a nation could belong to the other nation in his/her imagination. For example, of two neighbors, one's sense of belonging to one nation might be juxtaposed with the other's sense of belonging to a different nation. Gaps between nations have even greater influence on our personal neighboring with others.

Before September 11, I vaguely believed that in the future I could communicate with others without feeling any ideological and cultural differences because we had developed digital systems to share the same information globally. US capitalism and democracy seemed to gain huge influence over communist countries with the nonstop expansion of information networks. Mobile phones, email, and the Internet became strong tools to contact other nations. However, during the terror attacks and their aftermath, I realized globalization had transplanted nations' deep-rooted conflicts to our daily lives as well. I learned English in order to share the same linguistic ground with people who live all over the world, but when the disastrous accidents happened I found this common language and network also revealed our cultural, economic, and political gaps more seriously than before.

Globalization told me I could belong anywhere, but it actually meant I belonged nowhere. Our mobilization around the world had not created a new sense of belonging. In *Non-Places* (1995), Marc Augé argues that huge hours of today's lives are spent in anonymous places—supermarkets, airports, hotels, highway areas, computers, and cash machines. Augé's non-place, where people coexist as "distinct individualities, perceived as equivalent and unconnected,"[26] invades our life. What I confronted in Santa Cruz was this sense of non-place. I didn't belong anywhere. My communications with others were all temporal. Because of this uneasiness with my ongoing condition, I relied on the traditional idea of belonging formed in the nineteenth century—the idea of nations, communities, families, classes, genders, and even races.[27] Even if this idea of belonging did not work practically, I still felt these words had more substantial meaning for my sense of belonging. My sentiment for my home country was perceived as natural to the American people. Everyone at UCSC told me it was the best choice for my Japanese students and me to go home. Yes, to go home, to go back to my family, to return to my belongings would be the best choice. I realized I could not give any name to what I was creating at that time in Santa Cruz. Both they and I insisted that our belongings were different, and we could not name our personal communications. I sometimes wondered what I was doing in Santa Cruz. I did all things just to return home; however, was home the only place where I could find the reason for my existence? What was I creating with these people now? The world's temporal conditions and instabilities irritated me because I could not find any rightful purposes in this confused world.

It turned out that finding a flight home was easy, but its easiness made me mad. Several days after the terror attacks, I negotiated with my junior college and decided to cut our stay in the United States short and return home as soon as possible. I was afraid it might be difficult to get plane tickets on JAL (Japan Airlines) from San Francisco to Japan, thinking the flights might be full of Japanese passengers who also wanted to escape. Contrary to my expectation, however, just one week after the terror attack I found it was easy to find seats for all of our students. Almost all Japanese passengers had immediately evacuated the United States after September 11, and almost all visitors from Japan had cancelled their flights. All the JAL planes had plenty of seats, which annoyed me further. If, as before the attacks, JAL's seats were difficult to reserve for a large party within a week, and if the travel agency could only find one plane that could be reserved for us all, it would be easy for me to decide to buy tickets. Instead, the travel agent found I could choose any day and any plane to return to Japan. I now had to decide which day to cut our program short and make our return flight. It was still possible, even if the risk was low, that the plane would be hijacked or bombed. Which plane would be the best one for us? At that moment, I could not feel time went straight from beginning to end. The world was not fixed. It was not controlled by my will, and not completely decided by others. The future seemed to change on a whim by a combination of people's ideas and actions worldwide. I chose one flight from San Francisco to Tokyo without any special reason, hoping I might escape some other person's arbitrary malice.

We arrived at the San Francisco airport. It was the calmest airport in my life. No one wanted to visit airports. The announcements echoed in the hollow airport hall. I felt my American stay was empty, like this vacant airport. I met some people, but I couldn't give any meaning to my communications with them because I could not give them a proper name. The communications I had in the United States after the attacks were not as friendly—there were always some malfunctions. The unnaturalness of our existence suddenly appeared at the surface of all our behaviors, and this condition increased our communications' awkwardness. We realized we could no longer look at others without suspicion. This experience suggested the specific historical and geographical condition I was in as an international literature teacher in America. I asked myself again what kind of concept I needed to evaluate my temporal relationship with the world. Essentially, I do not have any innate motivation to have a relationship with America. I am one of the "others" in the English language, American literature in general, and the literature of the American South, even if I personally like William Faulkner's works; still, I feel my condition represents the historical position of all international literature teachers and ESL students. We are all foreigners to the studies of English and American literature, but we still try to explore these fields' issues as our own problems, hoping to have temporal, equal relationships with others according to these information materials. I questioned how I could evaluate these discursive activities.

When I returned to Japan with my students and I saw the faces of my wife and my students' families again, I felt we still needed something that we could agree on, something that could be applied to our everyday life, our everyday disasters—something that could be applicable everywhere. It could not be one of the old ideas—nation, race, religion, community, or family. It must be something possible even when alone in a foreign land, separated from family and belongings. This concept must be also useful for evaluating people's temporal relationships that are threatened by the huge pressure of the world's imperial impulse. Not only in times of terror, but also anywhere, anytime, in the real and imagined places, people make temporal relationships to create their own places in the world, and we still did not have a term to evaluate them. My exploration for a term to connect my personal experience with the world people's real and imagined experiences began on September 11, 2001. In the next section, I describe the process of finding the term, social combination.

FROM PERSONAL EXPERIENCES TO THE IDEA OF SOCIAL COMBINATION

It took thirteen years for me to develop the idea of social combination. In the process, I sometimes thought the family relationship was a basic human need for sustaining someone's trust in others, even if we often experienced troubles in our blood kinships. But, I thought again, if I created a new idea for today's personal relationship, I wanted it to be universal one that could include and support the various formations of families. My daughter was born while I was pondering the term I wanted, and I began to hope that my idea could be applicable to the world in which she would live.

I was still attracted to Faulkner's novels—*Soldier's Pay*, *Flags in the Dust*, *The Sound and the Fury*, *Light in August*, *Absalom, Absalom!*, and *Go Down, Moses*—because his early and middle works depicted each character's tragic isolation in a deeply stratified society. Faulkner's protagonists have trouble with their desires to belong to the families or communities because these traditional belongings do not function well anymore. It is revealed that the idea of kinships and fraternities does not work well in modern Southern society. The protagonists of these novels are looking to the past and demanding to regain romantic belonging or the traditional image of family and community, even if their demands seem perverted or nostalgic.

The narratives of these novels reveal that a fundamental idea to support the character's personal relationships—which cannot be bound by the idea of family or community—is needed for their region's reformation. However, Faulkner's early and middle works do not outline what kind of society Faulkner wants for the future. From *Soldier's Pay* to *Go Down Moses*, Faulkner criticizes the modern American South's desolate condition but does not explain how people can reconnect with each other in today's world. Donald Mahon in *Soldier's Pay* is the deeply wounded World War I veteran who waits only for death. Young Bayard Sartoris dies in an airplane accident in *Flags in the Dust*, satisfying his own sadistic demand for death. Quentin Compson in *The Sound and the Fury* is a suicidal, desperate young elite trying to keep his own incestuous desire for his sister in his isolated obsession. Popeye in *Sanctuary* is an impotent villain who only hopes for the ruin of the society. Joe Christmas in *Light in August* kills his white lover to revenge the community that abused him. Ike McCaslin in *Go Down, Moses* led an ideal life in the big woods when he was a child, and as an adult he discovers the cursed past of the South and rejects its inheritance, returning to the forest to live alone and retreat from the modern world. The protagonists of these works show Faulkner's severe criticisms of the modern Southern society, but they do not give any hints for the new social formation for the future.

In *Intruder in the Dust*, Faulkner depicts for the first time the community he demands. Faulkner insists socially marginalized characters' personal relationships can make the dialogical space in which people can keep their mobility and examine the community's problems from various points of view. Their personal experiences suggest the potential for an alternative social form in the American South.[28] *Intruder in the Dust* is not an ambitious novel when compared to Faulkner's masterpieces, *The Sound and the Fury*, *Absalom, Absalom!*, and *Go Down, Moses*; however, it suggests a future community in which individuals can personally make egalitarian relationships in a deeply stratified society. Specifically, the temporal, equal relationship between Chick and Lucas illustrates what Faulkner demands in his imagined world.

At that time, not only Faulkner's work but also real incidents in my country suggested a new concept was needed to recompose people's relationships in our world. Japanese people found they could be suddenly isolated from everything even in their own nation—in their own place—when the Tohoku earthquake and tsunami happened in Japan in 2011. Many lost their family members, and they had to create new, temporal relationships with others in their home place. Families were divided and individuals were alone in a disastrous time and place.

A new idea was needed for the people who were exiled from their home place, who did not have any kinship, and who could not share the same sense of place and community. I felt my concept also should be applicable to various formations of the people: the temporal gatherings in the emergency evacuation areas, extended families, same-sex couples, and parental bonds with their adopted children. Moreover, when I checked various works of literature of the American South, it seemed that not only Faulkner's fiction but also many other Southern literary works explored individuals' egalitarian relationships in various complex conditions. I wondered what kind of concept could evaluate all these personal relationships.

While I wondered over the years, digital technologies developed and people began to have smart phones, iPhones, and tablets. Skype gave us free international phone connections. Facebook and Twitter appeared, and people began to realize that both could be useful when disasters happened. People discovered personal broadcasts could be shared on those platforms, and learned that objective analyses were part of YouTube's international broadcast reports. When the Tohoku earthquake and tsunami happened on March 11, 2011 and demolished the Fukushima Daiichi Nuclear Power Plant, the Japanese government did not give detailed information on the damage of the power plants for some time, while the US broadcasts gave more objective, detailed analyses on the disaster.[29] These Web 2.0 platforms increased the frequency of people's short-term communications in common activities. These digital systems allowed us to have intermittent, personal connections with others. These temporal connections offered particular open platforms for all participants to exchange their utterances. My new term also should apply to these digital conditions. Since Paul Gilster's *Digital Literacy* (1997) claimed "interactivity, immediacy, and integration"[30] as the three key words of the web, digital media has progressed to allow multiple connections and immediate, private accesses. Web 2.0 has already achieved its interactivity and immediacy between individuals, but the problem is that we do not know our common goal in our private communications with others on the web.

I arrived at the idea of social combination in the summer of 2013 while I was searching for the proper words for my students' English textbooks. After my return from Santa Cruz, I changed my occupation from junior college teacher to university full-time lecturer studying and teaching American literature. Japanese university students' textbooks were different from the regular English textbooks—they used Japanese words to support the English text. In my students' textbooks, both languages were combined. I coined the term "Combined Education Edition" for the textbooks they were using, but I was not satisfied with my term yet. I wanted more general words that would represent Japanese ESL students' uneasy linguistic condition. Japanese ESL students often felt they could not master English completely—it's not their language, but the language of others. However, when I expanded the ESL students' frustration to our general linguistic experience, I found that no individual could completely possess and control language. Language is a phenomenon that occurs among individuals to combine their information and intelligence. I created the term social combination to represent individuals' efforts to have mutual communication with others in unsettled conditions. In our globalized world, people cannot rely on one culture, one language, and one nation's ideology to communicate with others. Individuals we meet have various cultures, languages, and nations. What we *can* do is to create the temporal common ground to make our conditions better with the others. It might be minimal common ground, but it might be what we need for our communications today. To reconfigure individual belonging in global relationships, social combination presupposes that people's efforts to make temporal, equal relationships are their social goals.

This term I partially hated, because it sounded cheap. It resembled an instant product like McDonald's combo menu, or sounded like an imitated term from Facebook's social network. Nevertheless, I accepted this term as the best term I could think of for today's globalized relationships because it could also be applied to every temporal relationship of an individual. Sometimes we are confronted with huge disasters we cannot predict. Sometimes we are separated and alone in foreign lands. Sometimes we are eternally parted. But we can still attain the egalitarian equalities in our personal relationships. I demanded social combination as the minimum moral in today's global condition after 9/11 and 3/11 in order to trust equality in relationships as our common ground for the future.

Individuals need global agreements in their private activities. In *Contesting Globalization* (2004), André C. Drainville argues that occupants of this planet should have "a common humanity"[31] as inhabitants who must share basic human solidarity and community and who should inevitably be together. After the development of Web 2.0, personal confrontation with other cultures' individuals is possible on the digital network. When individuals belong to different nations and cultures, every effort to attain a temporally equal relationship between them is important. Individuals should find a common ground in their communications to have appropriate discussions for a while. In this study, social combination is not a political procedure to decide something with majority voting, but a communication that sustains relationships for the better condition. Social combination does not demand any final decisions on discussions; rather, it demands to make better conditions for every participant as a historical process. According to Hirschkop, "Making a decision, in the sense of judging one perspective closer to the truth than others, is, in fact, not only unnecessary, but dangerous." Following Bakhtin's argument, Hirschkop claims the "complete victory" and "annihilation of the opponent" are just "rhetoric."[32] What we experience as a successful result of our dialogue is gaining a better condition for every participant. In contrast, our failures of discussion result in the deterioration of our common grounds or the advantage to one side of the participants. What I mean by individuals' equal relationships is a continuation of dialogue aiming to make our condition better. In literary works, also, equal relationships are realized in characters' personal dialogues, which do not give final judgments but give a better condition for each of them.

Even in the modern democratic nation, it is still difficult to attain equal relationships in individuals' personal interactions. As R. Barri Flowers reports in *Domestic Crimes, Family Violence and Child Abuse* (2000), peaceful, equal relationships have not been completely realized in our personal lives: "The reality is that in spite of laws that exist in every state making acts such as spouse battering, child abuse, child sexual abuse and elderly abuse a crime, family and domestic violence is a fact of life in the United States and elsewhere."[33] Personal violence is performed to make a definite hierarchy among individuals. Violence in our private life betrays that individuals have not completely found the method to build egalitarian relationships in their personal communications. Japan and other advanced nation's educators frequently find that bullying in students' personal relationships is serious.[34] In spite of the fact that Japan has attained economic success and now declares its high responsibility on the people's mutual sovereignty, serious personal violence persists in the family.[35] I expect the concept of social combination can be used to make social relationships among the members of the family more democratic. Temporal egalitarian relationships should be held between a parent and a child, between a husband and a wife, and between a brother and a sister to fulfill the affectionate relationships among family members.

The egalitarian agreement in personal relationships is difficult to attain in our usual conversations. As Yi-Fu Tuan suggests in "Introduction: Cosmos versus Hearth" (2001), today's life, especially its violent aspect, should be examined from multiple viewpoints that include our personal lives: "Verbal violence, either outright or in the subtler form of cruel put-downs, is not unknown in the family living room or bedroom; indeed, it may be more common there than in the boardroom or marketplace."[36] An individual needs an egalitarian relationship in his/her private life, even if he/she lives in a nation of democratic culture and politics. Education to create the mutual equality in our personal lives might be more needed on a global level. I began to search for a method to use digital networks to increase students' activities in more dialogic and equal ways. My experiences in class tell me that social combination is an effective idea to make students compose their dialogic relationships using their digital activities.

The concept of social combination is also applied to the pedagogy for ESL students who study the literature of the American South and use the digital platforms to support their class activities. When I read literary works of the American South with my students, I instruct students to find how some characters try to realize their own equal relationships. Also, in class, social combination is used to consciously make equal relationships among students both in class and on the digital network. In my pedagogical method, I demand to create more democratic conditions in classroom, because students tend to stick to the homogeneous small groups of classmates that have the same backgrounds. The idea of social combination resists this tendency and encourages students have discussions with various class members. After class, each student reflects on the dialogic activity in class and writes a paragraph combining his/her own words and the partner's words on the net. I would like to think these activities in class and on the net are not so-called "pair work" which tends to strengthen the students' social stratifications and cultural divisions in class, but "social combination" which insists upon building the people's ability to have the temporally democratic relationship with the others. In their personal combinations, students have short discussions to make equal relationships and fulfill their activities' purposes. On the digital network, students write a short paragraph that combines their opinion with that of others. I name these activities as social combination activities. In these activities, I hope students find plural viewpoints among themselves to compose their opinions in their readings and writings. Their efforts to have democratic relationships in class help students be aware of their intersubjectivity.

In order to make social combination as a globally useful term, this study defines intersubjectivity as the dialoguers' mutual agreement to make their conditions better in multiple situations. So far, as Hirschkop argues, intersubjectivity has been considered in the context of the development of Western culture. Quoting Jürgen Habermas's argument about intersubjectivity, Hirschkop claims Bakhtin idealistically uses the term dialogism as the "modern intersubjectivity embodied in the form of the European novel."[37] He means that Bakhtin's intersubjectivity in dialogue has been the historically and geographically circumscribed common plane for people in the Western enlightenment. Also, Habermas, the German philosopher and social theorist, argues this intersubjectivity should be the basic presupposition for today's usual communication among people who have different opinions. His idea of intersubjectivity is deeply soaked in the idea of Western reasoning and discussion. In *The Theory of Communicative Action*, Habermas deduces that, trusting the long tradition of Western rationality and its reality, people can presuppose the same abstract plane, which is freed of all specific content:

> In communicative action we today proceed from those *formal presuppositions of intersubjectivity* that are necessary if we are to be able to refer to something in the one objective world, identical for all observers, or to something in our intersubjectively shared social world. The claims to propositional truth or normative rightness actualize these presuppositions of commonality for particular utterances. . . . Validity claims are in principle open to criticism because they are based on formal world-concepts. They presuppose a world that is identical for *all possible* observers, or a world intersubjectively shared *by members*, and they do so in an abstract form freed of all specific content. Such claims call for the rational response of a partner in communication.[38]

Habermas insists that truth is just a type of validity claim, and someone's truth—someone's validity claim—should be examined from the various points of view in the rational discussion with the other members. Yet, his argument seems to believe that all participants can preliminarily share a rational consensus. This presupposition of people's agreement on human rationality that can be held "in an abstract form freed of all specific content" overlooks its alienating other cultures that may not accept the European reasoning and its way of arguing. His trust in the enlightened commonality for the rationally unified consensus should be reconsidered in today's conditions.

As Hirschkop argues, in both cases of Bakhtin and Habermas, "the subject is asked to abstract its action from any consideration of its life-story and the narrative of its social context," but now what we need is a more direct sense of being with others when we have a narrative, because "for the modern personality the desire to act justly is inextricable from the desire to be a righteous or just person, to live in a just community, perhaps to fight for a just future, which is to say that it cannot comprehend moral action without thinking of it as a moment of a narrative." In the same way as a blogger interacts and perceives the global audience's reaction in his/her narratives on the web, today's people should feel their words' moralistic correctness/failure in their personal recognitions of their dialoguers' reactions. To keep intersubjectivity in dialogue, our globalized condition requires us to incessantly reform our common ground and adjust our equality in our heterogeneous combinations. Hirschkop suggests today's individuals need to recognize their mutual differences in order to include intersubjectivity in their own narratives. He says it is our desire "to live in a just community," which means that, in today's globalized world, the idea of intersubjectivity should be empowered not as a particular culture's common rule, but as actions that create new common grounds for culturally different individuals.[39]

A COMMON PURPOSE IN GLOBAL COMMUNICATIONS: AN EPISODE AND ANALYSIS

What kind of common purpose can be found in today's expanded relationships? In today's globalized condition, one particular area's common purpose is not necessarily beneficial for another area's people. We already know that improvement in life cannot be fully realized by the principle of Western enlightenment. As Gergen argues, instead of the modernist worldview that values enlightenment through reason and observation by which "humans can discover the fundamental essences of the universe," we find "an open slate emerges on which persons may inscribe, erase, and rewrite their identities as the ever-shifting, ever-expanding, and incoherent network of relationships invites or permits."[40] In this condition, what is today's common purpose that can be shared with everyone? Is it possible for us to find a common phase that can evaluate various activities in the global space?

It is difficult to give a satisfying answer to this question, but here I outline my own personal experience—a conversation with my daughter in Spring 2014—to demonstrate how individuals' personal exploration can be connected with today's common purpose. My daughter was ten years old at the time, and she loved books. She expanded her worldview every day and developed interests in social and scientific knowledge. She began to think about the beginning and the end of the universe, and enjoyed scientific columns in kids' magazines. One day, my daughter came to my room and asked what would happen with the end of the Sun. I told her that in its final stage the Sun would become a red dwarf star, which would expand to swallow the Earth. She asked what would happen to the people.

"It will not happen so soon. It takes a long, very long time—maybe more than ten billion years—for the Sun to swell up to that level. It's a story in the very distant future, but, anyway, I think human beings will already be extinct at that time." I said.

"Human beings will die out?" She looked shocked.

"I'm not sure. When I was a child, some science books for children told me that human beings can make big spaceships, and they will fly off from the Earth to search for other solar system's planets. But can you imagine that such long distance space travel in massive ships will be possible? I don't think so. Someday, the Sun will swell up to its limit, and then begin to shrink again, losing its energy. At that stage, I assume, human beings will have disappeared from this universe."

"So the Earth and human beings are not eternal." My daughter sadly looked around my room. Maybe she had already perceived that some end would come to the Earth and its people, but she was now shocked because she confronted the scientifically proven end of our society. She seemed to be feeling a strong sense of helplessness. I could understand her despair. When I was almost my daughter's age, I also felt the same helplessness reading the scientific explanation about the end of our solar system. I felt a strong despair about humanity's future, although my science magazine for kids showed me humans' great escape to the universe with a vivid cartoon. I felt disappointment when I saw that cartoon—I knew such an optimistic plan was an adult fantasy. Maybe I felt disappointed with the adults who could tell such a big lie without any hesitation. "I don't want to be a member of such adults," I thought at that time. And now I wanted to be an adult who could give a serious answer to my daughter's sad feeling. We were silent for a while.

"Then, why are we doing all these things now?" My daughter suddenly asked me, spreading her hands around. I understood what my daughter was asking. Why are we all creating these highly developed societies for such a desperate future? Why are we doing all these things? That's exactly what I asked myself when I was a child. Why do we demand more development of our society if we know that the inevitable extinction of all these cultures will come in the future? Now I had to answer this question as her parent—or, I had to answer this question for my own sake. But I didn't feel that I could properly answer the question at that time, so I told her to wait for a while. I would surely answer it tomorrow.

The next day, I dined with my daughter. My wife had gone to her evening PTA gathering, so it was just my daughter and I eating our Japanese meal at the kitchen table. While I was eating my steamed rice, I asked my daughter if she remembered the previous day's question. "Yes Dad," she said while she ate.

"OK. I couldn't give you an answer to your question yesterday, but now I want to tell you my idea—my answer—to your question. Someday, the Sun and Earth will disappear from this world. Someday, human beings will become extinct. All the cultures on this planet will be replaced. Yes. But imagine, such disappearances happen every day. Someday, your father will die. I will disappear from this world. It's true, right?"

"It's true," she said. She was now eating sashimi my wife had cut for us.

"Yes, it's true. Your father knows that his life will end someday. But I will not lose any hope in building up all these things, all our life." I spread my hands around like my daughter had that previous day, holding a rice bowl in my left hand and chopsticks in my right. "No. I don't stop my effort to make our life happy and hopeful, here and now. You know what I mean? We are building up our life for something we can share. Not only for me, not only for you, but we are now creating something between us. And remember, even if I die tomorrow, I will not cease my effort today. Because this is our happiness; this is our life. I believe that even if we all die, someone, or even something, will continue this invisible thing again and again. You know, this world among us continues, even if both of us die. We are living now to improve this invisible thing."

I pointed to the space between my daughter and me with my chopsticks. She was vaguely looking at the space between us and said "Yeah." I did not know whether she understood what I said. Maybe I would never know. Maybe she created her own opinion after hearing my words. But I believed the idea that was created among us would continue. I hoped it would continue even after we all died out. I looked at my daughter's face. Tomorrow, a terror attack may come. A big earthquake may happen. The nuclear plant near our area may explode. Yet we will continue our ordinary life working to create this invisible thing today to make this universe better.

My dialogue with my daughter suggested to me that once one's recognition of the world is shared with the other, it inevitably supports the common purpose of the world. Intersubjectivity is meaningful, not because it is a common platform for Western enlightenment, but because it can suggest individuals' abilities to communicate with others to build up their social grounds and imagine their better conditions. Intersubjectivity is a key term to connect personal actions with public purposes. Once personal dialogue is narrated in public, it inevitably contains the common values that are shared among participants. Sometimes this intersubjective effect in dialogue increases negative tendencies in both of the participants and sometimes it increases positive tendencies. In both cases, however, we can find the place that is shared between the participants' particular contexts in order to consider the world between individuals. This temporal, common place is where miscellaneous values and worldviews struggle and survive. Neither my daughter nor I can escape this field. This condition suggests our moral recognitions and metaphorical ideas essentially come from our efforts to continue personal dialogue with others to make a common ground—two or three people agree to define one idea, and then the conclusion of this personal dialogue expands to the world as a discursive common settlement for all. Both Plato's and Confucius's ethical arguments tell us that people's wisdom and morality come from their personal dialogues. What I want in my literary class is students' experience of this budding condition to invent commonality from their personal dialogues.

Faulkner expresses this personal combination as "the combined intelligences."[41] The world's combined intelligences see their future in their present condition to make a common ground. When Faulkner was asked in a public interview at the University of Virginia in 1957 why he feels "it is necessary to bring in" the future while he is "telling the present,"[42] Faulkner explains that individuals' incessant intellectual struggles to create their social formation with others do not change from the past to the future:

> Well, a man's future is inherent in that man—I—in the sense that life, A.D. 1957, is not the end of life, that there'll be a 2057. That we assume that. There may not be, but we assume that. And in man, in man's behavior today is nineteen fifty—two thousand and fifty-seven, if we just had a machine that could project ahead and could capture that, that machine could isolate and freeze a picture, an image, of what man will be doing in 2057, just as the machine might capture and fix the light rays showing what he was doing in B.C. 28. That is, that's the mystical belief that there is no such thing as *was*. That time *is*, and if there's no such thing as *was*, then there is no such thing as *will be*. That time is not a fixed condition, time is in a way the sum of the combined intelligences of all men who breathe at that moment.[43]

In Faulkner's imagination, people are incessantly building up their daily communications for their intellectual formations. People's struggles to make social connections with each other do not change from the past to the future. Faulkner recognizes that this series of different social formations is—even if it is a series of failures—human history.

I teach both Fa(u)lkners' works as part of people's process to explore the better conditions of social relationships in the American South. In W.C. Falkner's romantic world, characters always identify their ideal goals as eternal ones. The unity of the hero and the heroine in a narrow community is the ultimate goal of W.C. Falkner's romantic works. This romanticized pattern makes W.C. Falkner's literary work an ahistorical repetition. In contrast, in Faulkner's imagination, characters demand to create their own social formations to make their conditions better, even if there are some limitations in their abilities. From Faulkner's perspective, human history is a continuous process of failures to establish people's ideal social forms. As a counterpart of the main characters' romantic unities in W.C. Falkner's works, characters' social combinations are key personal conditions of Faulkner's works.

Faulkner's great achievement is, of course, his creation of a unique county—Yoknapatawpha—and its history. Faulkner settles Jefferson as the central town of this imagined county in Mississippi and creates this imagined place's history and various people's lives as a process of efforts to have their ideal communities. It is a record of the people's failures to create their personally socialized worlds. Some characters follow the cultural tradition of the American South and try to create their family-oriented communities. Some characters step out of these family-oriented obsessions and try to create some other communities. In each case, Faulkner implies that the characters' formation of their communities has limitations—that is to say, they form their communities for a while, but then their efforts come to an end, revealing their designs' fatal flaws. This pattern also appears in the characters' short formations of social combinations. Individuals form social combinations for a while, but their combinations do not continue eternally because of their historical and geographical limitations.

In Faulkner's fiction, the history of the American South appears as a series of people's trials and errors to make their communities. His perspective is large enough to contain W.C. Falkner's life and his literary works as well. W.C. Falkner insists on a romantic pattern of successful life, both in his life and his literary works, and Faulkner's fiction critically examines this pattern as the curse of his fictional world using Yoknapatawpha County. And then Faulkner's later fiction—after *Intruder in the Dust*—reveals that Faulkner finally overcomes W.C. Falkner's ideal life design and his literary works, accusing them from the viewpoints of socially marginalized people. Faulkner's works suggest people's power to continue their efforts to form alternative social formations despite their failures. Especially after *Intruder in the Dust*, Faulkner's characters keep on forming social combinations in their personal activities, showing the American South as a place where people can hold personal relationships sharing some common living sensations. Faulkner's ambitious vision includes the whole history of the Southerners and even the history of human beings, who demand to continue their lives with others.

THE AMERICAN SOUTH AS A CULTURAL ENTITY

Some current scholars who argue American literature in a global perspective tend to see Faulkner's literary world as an enclosed, narrow place to which his characters inherently relate. For example, in *The Global Remapping of American Literature* (2011), Giles argues, "For Faulkner, the South is not so much a region in the political or cultural sense but precisely the kind of soil or 'territory' to which his characters relate by instinct."[44] However, when a reader oversees Faulkner's works in their relationships with W.C. Falkner's, one can find that Faulkner's literature discloses the local characters' instinctive connections to reveal that their relations are personal illusions. Some family members believe that their kinship ties generate instinctively, but some others divulge that such local and family ties are always in danger of breaking down and their condition is just a part of the broad process of the changing formation of the society. Faulkner sees the development of human history not as the consecutive lineages, but as a clumsily composite sequence with personal connections and struggles.

Even if some family members feel that they are inherently cursed by one fatal pattern, Faulkner's imagination historicizes how such a local curse has been caused by some specific family member's obsessive design. In his early and middle works, Faulkner, who is personally caught by his great-grandfather's romantic gravity, analyzes how various communities' and families' so-called instinctive connections originally come from some particular characters' discursive imaginations that effectively work to control some others' fates and relationships. Then he shows, especially in his later works, that characters can develop the ways to have temporally equal relationships even if they are individually captured by different monologic worldviews. This attitude can be broadly understandable for today's international readers. In worldly related personal relationships, the international readers commonly see that diversified individuals still variously gravitate to monologic imaginations, religions, and ideologies—some may be attracted to Anglo-Saxon traditions, some to Sinocentrism, some to Judaism, some to Islam, some to various nationalisms, and some to multiple secessionisms—and yet these individuals are also digitally connected and personally giving efforts to have equal relationships among them. ESL students also find that people are in such controversial conditions when they try to create their social formations with their worldly expanded dialoguers.

In spite of particular families' and communities' instinctive ties, an individual's specific conduct gives centrifugal effects to people's relations. This perspective is useful to give meaningful hints to the international reader's dialogical dealing with a particular culture and its appearance. A dialogical relationship with the other culture is still difficult for everyone because the world is not flat and it is always threatened and gravitated by particular monologic imagination. Even if, as Thomas L. Friedman writes in the opening of *The World Is Flat*, an American author finds that some Indians in India have taken "American names" and speak English imitating "American accents,"[45] it does not mean the world has flatly shared the universal rule; it just means many parts of the world have been dominated by the American culture and language, and the Indians temporarily choose this particular mode to communicate with the others in their efforts to have an equal relationship. The problem is that many people inside and outside of America tend to consider this way of communication as the realization of the common rule, or that they do not fully recognize that this American mode is a temporal one. It is difficult for them to historicize this mode, because digitally connected people tend to internalize the American communication and the American cause naturally. Especially for the ESL student, it is quite hard to deal with this internalized Americanness when they want to have truly dialogic relationships with American culture. The term "the American South" is useful to suggest that this problem of internalized Americanness, which is now expanding to the world, has historically occurred in the local area of America.

Some recent scholars, who prefer to insist upon the multicultural condition of the United States, may want to use "U.S. South" as the more suitable term to recognize this area as a postcolonial space of the imperial United States and focus on this area's circum-Caribbean culture. In *Look Away!*, Jon Smith and Deborah Cohn consciously use this term "U.S. South" to acknowledge "its postcolonial status without ignoring the region's signal histories of oppression, its cultural specificity, *and* its cultural diversity" and to concentrate on "the Deep South not as a unified or imagined community but as a scene of the cultural conflicts that white imaginings of community seek to forget."[46] This attitude is quite attractive. If I want to focus on, for example, New Orleans and its Creole culture, this term might be the precise one to use; however, I do not use the term "U.S. South" in my argument because this term may avert my argument's focus from two Fa(u)lkners' imagined American South to the Caribbean-oriented multiculturalism.

The term "the American South" insinuates that two Fa(u)lkners' area belongs to the American culture, and that the matter of the American South is the problem of American people—or, of the people who internalize this region's culturally required condition. I try to reopen this condition, to reconsider the problem of "the American South" and the history of the imagined "Southern" issues as the matter of the people in the world. This also means that, when I use the term "the literature of the American South," I critically inherit the usage of this term in *The Cambridge Companion to the Literature of the American South* and historicize the traditional terms, "Southern literature" and "the South." One of the contributors of *The Cambridge Companion*, Matthews developed Kreyling's argument to observe "Southern literature" and "the South" with some irony. Matthew argues as follows: Using these traditional terms, "Southern literature" and "the South" in *Inventing Southern Literature*, Kreyling analyzes that, even when the Southern critics around the end of the twentieth century found the meaninglessness of the region's literary lamentation on its region's tragic cause and its fate, they still relied on "Faulkner's shame" on his province's misbegotten history to sustain their "modern Southernness" and showed their "melancholic" sympathy with the Quentin-Compson-sort intelligence.[47] I develop Matthew's argument and ironically use capitalized words, "Southern" and "South," in my argument to suggest that Faulkner's modern Southern sentiment effectively works even for today's readers to internalize this region's painful otherness as "American" one. Because of the white-elite's traumatized attitude, even international readers can imagine that this area is something mythic, something foreign, something strange, but at the same time can perceive that it strongly inspires their desire to recompose their imagination to unify this otherness as their internalized America. Using the term, "the literature of the American South," my argument targets this "American" condition of "the South," which implies that this literary region still belongs to the monologic American culture and ultimately inspires readers' desire to identify themselves with the Americans who cope with the Southerner's white-oriented

"American" cause, and obstructs (but potentially encourages) readers' dialogic relationships with this region's cultural entity. Faulkner's later works are the records of a modern author's self-denial resistance against this internalized tendency, and his characters' alternative relationships are suggested even when they are gravitated to a particular culture's centripetal force.

TWO FA(U)LKNERS' LITERARY WORKS AND ESL STUDENTS

I use both Fa(u)lkners' literary works as ESL students' texts for the following three reasons. First, W.C. Falkner's works are good texts for helping ESL students understand the typical romantic pattern of the nineteenth century's popular tales, as well as its ideological shortcomings. ESL students enjoy W.C. Falkner's stories as entertainment, but at the same time they can find the problems of his self-centered worldview. In W.C. Falkner's works, the romantic heroes interpret liberty as the will to protect their helpless, weak people from enemies, but it turns out to be the Anglo-Saxon heroes' imperial impulse to dominant the marginalized. Because of their discovery of the shortcomings of W.C. Falkner's romantic works, ESL students understand the reason why the modernist author, William Faulkner, had to develop new forms to make his novels resist the romantic worldview in the nineteenth century. Second, Faulkner's portrayal of human history is appropriate for ESL students. Faulkner feels human history is a series of people's failures to form ideal communities. So far, there has been no successful formation of the ideal community, but Faulkner's works suggest that people can continue their efforts to explore new formations of society in various ways. This historical perspective is an appropriate social model for today's people, who should continue to explore their new formation of society in their globalized condition. Third, these two authors' literary works show a historical development from romantic monologism to the modern dialogical formation of individuals. This pattern also suggests that the modern novel's dialogism, which presupposes every voice's multiple links and interrelationships, develops to overcome romantic monologism, in which a heroic voice ultimately dominates the other voices. Readers can understand this historical development by following the various forms of personal combinations in these two authors' works. Faulkner strictly criticizes the American South's romantic illusion and instead offers a democratic relationship of individuals—I call this relationship "social combination" in this study—in *Intruder in the Dust*. Faulkner's depiction of individuals' relationships in *Intruder in the Dust* gives a suggestive social form for today's individuals because he focuses on people's capacity to form democratic relationships among individuals who live with an expanded social mobility. I also apply this idea of social combination to guide ESL students who have various relationships in today's digitally expanded environment. ESL students who must learn digital literacy in today's culture will study the historical development of individuals' relationships in both Fa(u)lkners' works, which may hint at their social formations in the future.

In the following three chapters, I explore the ways to apply the idea of social combination to interpret and teach both Fa(u)lkners' works in ESL classes that are equipped with digital devices. Chapter 1 examines W.C. Falkner's works—*The Siege of Monterey*, *The Spanish Heroine*, *The White Rose of Memphis*, and *The Little Brick Church*—in order to find their thematic connection with Faulkner's works. W.C. Falkner romantically naturalizes the white male hero's dominance of his narrow community's members. Faulkner's early and middle works critically examine how modern white males in Yoknapatawpha County are attracted to the imperial impulse present in W.C. Falkner's works. Faulkner's early and middle works re-accentuate W.C. Falkner's romantic motifs, and yet they still illustrate how the modern author is obsessed to the Southern romantic hero's self-centered world. Faulkner finally overcomes his great-grandfather's romantic worldview in *Intruder in the Dust*, in which Chick Mallison makes temporal communications with the socially marginalized people to resist the white males' dominant discourse. These two authors' literary forms are argued as a historical development from the romantic author's monologism to the modern author's dialogism. Especially in *The Little Brick Church*, W.C. Falkner monologically depicts the hero's imperial attitude toward the weak and helpless as the essence of America's liberal spirit. He suggests that the plantation culture and paternalism in the American South is a direct descendant of the original social structure of the pre-Revolutionary American culture. Faulkner criticizes this theme when he depicts Gavin Stevens's paternal attitude toward African Americans and women in *Intruder in the Dust*. The monologism in W.C. Falkner's works is one of the main objects of Faulkner's criticism.

In chapter 2, Faulkner's *Intruder in the Dust* is used to study how the characters' social combinations appear as the modern individuals' efforts to have democratic relationships in a deeply stratified society. An elderly black man, Lucas Beauchamp, is wrongly jailed and accused of the murder of a poor white man. While being threatened by the white locals' intention to lynch him, Lucas hopes a young white boy, Chick Mallison, can investigate the murder and prove his innocence. His request motivates Chick to see the case from Lucas's point of view. The mob's sadistic desire for lynching betrays the falseness of the white community's moral superiority and unity. In order to challenge their cultural and institutional constraints, Chick cooperates with a young African American friend, Aleck Sander, and an old white spinster, Miss Eunice Habersham. These marginalized characters' collaboration to stop the lynching proves that the dialogues between individuals from different social strata can find an appropriate solution as they examine an issue from various points of view. These socially marginalized characters, who are less anchored to the fixed positions in the community and less constrained by their community's institutional norms, are the local agents who can make temporal deals with others to gain what they really want. Chick finds that the basic structure of society is not the fixed social positions of individuals but the individuals' possibility of interactive exchanges on common ground. In *Intruder in the Dust*, Faulkner reconfigures the personal dialogical space, not the place in which the individual confronts the ghost of romantic unity, but the place in which they can create dialogue with others. In this story, the characters' socially stratified conditions—based on the ideological concepts of race, class, and gender—are depicted as particular conditions under which new common ground can be made, and as a metaphor for the world in which we struggle to create dialogic space.

This study's pedagogical project is reported in chapter 3. Today, W.C. Falkner's works belong to the public domain—most of his works can be internationally read on the Internet. Faulkner's modern works are translated into various languages and have gained worldwide readers, and they will soon belong to the public domain as well. This situation is a strong contrast to the twentieth century's paper-based publishing. In the twentieth century, the nineteenth-century popular Southern novels were not republished and they were difficult to study in class. These works were considered minor and worthless, and they were almost neglected in academic arguments. However, this has changed in today's digital age. Now, thanks to various digital archives of the nineteenth century's minor literature, it is possible for teachers to study these minor works in class if they combine such texts with more popular texts in order to explore their dialogical relations. It has been difficult to establish the proper method to teach nineteenth-century Southern works in undergraduate courses; however, in this study I explore a pedagogical method to teach ESL classes to read and discuss nineteenth-century minor works and the other major works of American literature using digital devices. I cooperated with Seinan Gakuin University in Japan to create a Moodle database in which students could combine their own opinions with someone else's to create a paragraph after they have discussed in class how to interpret both Fa(u)lkners' works by themselves. Connecting students' class discussions with their writings on Moodle, I required students to write their opinions and include other class members' points of view. This project was researched in two literary classes. In one class, I taught William Faulkner's short story, "Wash," and W.C. Falkner's short novel, *The Spanish Heroine*. In the other class, I taught William Faulkner's *Intruder in the Dust* and W.C. Falkner's *The Little Brick Church*. Comparing both authors' works, I instructed students to recognize how these two authors express their characters' relationships. Students had social combination activities to discuss with their partner and write a paragraph on Moodle that reflected their arguments in class. Using the concept of social combination, ESL students made efforts to have personal, democratic relationships with others in order to improve their understanding of both Fa(u)lkners' texts and today's digital literacy.

The concept of social combination is effective to connect both Fa(u)lkners' works with ESL students who use digital devices and systems in their everyday life. In W.C. Falkner's works, the narrow community is imagined as one in which the white males finally gain their dominant positions. In contrast, Faulkner's works criticize the complete unity of the narrow community under the dominance of the white males, ironically using W.C. Falkner's romantic motifs. Especially in his later works—beginning with *Intruder in the Dust*—Faulkner reconfigures his imagined community as a place in which the marginalized characters' social combinations reexamine the white male dominance from various points of view. Today's readers, who expand their personal relationships in digital networks, can find their problem of social formation has already been predicted in the process of development from W.C. Falkner's works to Faulkner's.

NOTES

*Throughout the book, italics in quotes are in the original.

1. William Faulkner, *Faulkner in the University* ed. Frederick L. Gwynn and Joseph L. Blotner (1959; Charlottesville: University Press of Virginia, 1995), 139.
2. Sharon Monteith. "Introduction: Mapping the Figurative South," in *The Cambridge Companion to the Literature of the American South*, ed. Sharon Monteith (Cambridge: Cambridge University Press, 2013), 1, 9.
3. Tara McPherson, *Reconstructing Dixie: Race, Gender, and Nostalgia in the Imagined South*. (Durham: Duke University Press, 2003), 1.
4. Gwen Solomon and Lynne Schrum write that the term Web 2.0 was coined in 2004 to encompass "the growing collection of new and emerging Web-based tools." Many of these Web 2.0 tools are "free and available to all" and "social in nature and promote self-expression, such as the community networks, blogs, wikis, and photo and video sharing sites." Yet, the budding development of an interactive digital network can be found much earlier. Tara Brabazon argues that 2001 was "a key period of transition between web 1.0 and web 2.0"; according to her, up until 2001 most online activities were used for "searching, reading and viewing, rather than commenting, writing and uploading," but now "the read write web" has "entered popular cultural currency." See Gwen Solomon et al., *Web 2.0: New Tools, New Schools* (Eugene: International Society for Technology in Education, 2007), 13; Tara Brabazon,

"Introduction: New Imaginings," in *Digital Dialogues and Community 2.0: After Avatars, Trolls and Puppets*, ed. Tara Brabazon (Oxford: Chandos, 2012), 2.

5. Allen Tate, "Remarks on the Southern Religion," in *I'll Take My Stand*, by Twelve Southerners (1930; Baton Rouge: Louisiana State University Press, 2006), 171.

6. W.J. Cash, *The Mind of the South* (1941; New York: Vintage, 1991), 32.

7. Michael Kreyling, *Inventing Southern Literature* (Jackson: University Press of Mississippi, 1998), 6.

8. Patricia Yaeger, *Dirt and Desire: Reconstructing Southern Women's Writing, 1930-1990* (Chicago: University of Chicago Press, 2000), 38.

9. Richard Gray, "Foreword: Inventing Communities, Imagining Places: Some Thoughts on Southern Self-Fashioning," in *South to a New Place: Region, Literature, Culture*, ed. Suzanne W. Jones and Sharon Monteith (Baton Rouge: Louisiana State University Press, 2002), xxi.

10. Kreyling, *Inventing Southern Literature*, 6.

11. Cash, *The Mind of the South*, 31.

12. Hubert J.M. Hermans and Harry J. G. Kempen argue that the self feels its own social position changes according to its relationship with others. They explain this shift of the self by critically examining G. H. Mead's assertion: "the self appears in experience essentially as a 'me' with the organization of the community to which it belongs." Hermans and Kempen are not satisfied with Mead's analysis, and offer the concept "*I position*" in order to incorporate the self's frequent shift in its relationships with others. However, Mead's theory is still attractive for literary study. See G.H. Mead, *Mind, Self, and Society* (Chicago: University of Chicago Press, 1934), 200; Hubert J.M. Hermans et al., *The Dialogical Self: Meaning as Movement* (San Diego: Academic Press, 1993), 109, 115.

13. I owe part of this argument to Michael Kreyling's comment on Barbara Ladd's *Nationalism and the Color Line in George W. Cable, Mark Twain, and William Faulkner*. See Kreyling, *Inventing Southern Literature*, x.

14. Since Ralph Ellison compared Mark Twain's *Adventures of Huckleberry Finn* with William Faulkner's *Intruder in the Dust* in his essay, "Change the Joke and Slip the Yoke," in 1958, these two works have been coupled as typical works that depict the relationship between African American and white youth in the American South. See Ralph Ellison, "Change the Joke and Slip the Yoke," in *The Collected Essays of Ralph Ellison* (New York: Modern Library, 1995), 104–5.

15. Joseph R. Millichap, *Dixie Limited: Railroads, Culture, and the Southern Renaissance* (Lexington: University Press of Kentucky, 2002), 9, 25.

16. Taylor Hagood, *Faulkner's Imperialism: Space, Place, and the Materiality of Myth* (Baton Rouge: Louisiana State University Press, 2008), 91, 30.

17. In W.C. Falkner's case, scholars most often mention *The White Rose of Memphis* as his best work. Donald Philip Duclos writes *The White Rose of Memphis* should be placed "at the head of Colonel Falkner's literary works." In Faulkner's case, many scholars consider *Absalom, Absalom!* as his best work. Cleanth Brooks argues *Absalom, Absalom!* is "the greatest of Faulkner's novels." See Donald Philip Duclos, *Son of Sorrow: The Life, Works and Influence of Colonel William C. Falkner, 1825-1889* (San Francisco: International Scholars Publications, 1998), 162; Cleanth Brooks,

William Faulkner: The Yoknapatawpha Country (1963; Baton Rouge: Louisiana State University Press, 1990), 295.

18. Mikhail Bakhtin's term, heteroglossia, and his argument can still be useful to analyze the globally expanded condition. In the end of postmodern era, around 1990, some scholars used Bakhtin's heteroglossia and his argument to explain their expanding experiences. In his social analysis, *The Saturated Self* (1991), Kenneth J. Gergen uses "the term *heteroglossia*" to argue that "the existing language of any culture bears the remnants and admixtures of languages from various subcultures and historical eras," and that "postmodernism invites a heteroglossia of being, a living out of the multiplicity of voices within the sphere of human possibility." The postmodernist perspective gives equal value to the romantic and the modern traditions, and furthermore opens its view to incorporate or borrow other cultural forms such as "Asian, Polynesian, Arab, Indian, and African cultures." Thanks to the expansion of information technology, many postmodern works are permeated in multicultural ways. Gergen, who feels technological development has gone beyond the control of individual capacity and the self is saturated by the overflowing information, argues Bakhtin's model is useful to positively analyze an individual who is part of the broader world. In "Postmodernism and the Human Sciences" (1992), Patti Lather also uses Bakhtin's idea to question objectivity in the psychological relationship. Insisting that our attention should be turned away from efforts to represent things to what Bakhtin terms "the framing authorial context" (DN 358), Lather argues that post-structuralism regards research as "an enactment of power relations" and it focuses "a mutual, dialogic production of a multi-voice, multi-centered discourse." In this condition, objectivity "is seen as textual construction more fruitfully displaced by a deconstructive emphasis on writing as an enactment of the social relations that produce the research itself." As a postmodernist who questions the intellectual framework of the dominant culture, Lather insists on "a decentred subjectivity." These scholars insisted that participation in the world requires the decentering of every individual and the recognition of the social language's heteroglossia. See Kenneth J. Gergen, *The Saturated Self: Dilemmas of Identity in Contemporary Life* (1991; New York: Basic Books, 2000), 247–48; Patti Lather, "Postmodernism and the Human Sciences," in *Psychology and Postmodernism*, ed. Steinar Kvale (London: Sage Publications, 1992), 97; Steiner Kvale, "Introduction: From the Archaeology of the Psyche to the Architecture of Cultural Landscapes," in Kvale, *Psychology and Postmodernism*, 4.

19. J. Hillis Miller, *On Literature* (London: Routledge, 2002), 3.

20. Tzvetan Todorov, *Mikhail Bakhtin: The Dialogical Principle*, trans. Wlad Godzich (1981; Minneapolis: University of Minnesota Press, 1984), 104.

21. Ken Hirschkop, *Mikhail Bakhtin: An Aesthetic for Democracy* (Oxford: Oxford University Press, 1999), ix.

22. Amartya Sen, *Identity and Violence: The Illusion of Destiny* (New York: W.W. Norton, 2006), 29.

23. Hubert J.M. Hermans and Harry J.G. Kempen, *The Dialogical Self: Meaning as Movement* (San Diego: Academic Press, 1993), 49–50.

24. Flaubert's words are quoted in Hermans and Kempen, *The Dialogical Self*, 50.

25. Michel Foucault, *The Archaeology of Knowledge and the Discourse on Language*, trans. A.M. Sheridan Smith (New York: Vintage, 2010), 155.

26. Marc Augé, *Non-Places: Introduction to an Anthropology of Supermodernity*, trans. John Howe (London: Verso, 2000), 111.

27. Benedict Anderson's argument in *Imagined Community* suggests that the modern sense of belonging, which began with the Euro-American imperialism in the nineteenth century, has left a serious stigma in the modern identity. With the development of the modern imperial nations, people have used their ideas of belongings—the idea of nations, communities, families, classes, genders, and races—to naturalize their senses of dominance against the others. Therefore, applying Anderson's logic, Barbara Ladd argues, in *Nationalism and the Color Line in George W. Cable, Mark Twain, and William Faulkner*, that "the concurrent rise of U.S. imperialism abroad and radical racism inside the United States, which held that blacks, freed from the civilizing influences of slavery, were retrogressing into savagery and would eventually die out in the competition with whites, was no coincidence." If a modern person feels that belonging to the particular group or category generates the sense of safety, it means that such a naturalized sense of identity is inevitably connected with the power of imperialism and it is quite difficult for the modern individual to escape from this historically composed structure. See Barbara Ladd, *Nationalism and the Color Line in George W. Cable, Mark Twain, and William Faulkner* (Baton Rouge: Louisiana State University Press, 1996), 148.

28. Characters' attempts to have temporal, equal relationships between individuals in Faulkner's *Intruder in the Dust* shares its prophetic quality with Martin Luther King Jr.'s words and actions in the civil rights era. At the conclusion part of his "I Have a Dream" speech in Washington, D.C., on August 28, 1963, King symbolically use the temporal, equal connection between the African American and the White to suggest an ideal goal of people's democratic relationship: "I have a dream that one day on the red hills of Georgia, the sons of former slaves and the sons of former slave owners will be able to sit down together at the table of brotherhood." In his actions, he calls for the individual's practical attempts to realize this ideal goal in so-called lunch counter sit-ins and other nonviolent demonstrations. In King's words and actions, the temporally egalitarian relationship between individuals in the American South is approved as an effective endeavor to connect people's daily actions with the goal of democratic social reformation. See Martin Luther King Jr., "'I Have a Dream': Speech Delivered at the March on Washington, August 28, 1963," appendix to *King's Dream*, by Eric J. Sundquist (New Haven: Yale University Press, 2009), 232.

29. As Akira Nakamura and Masao Kikuchi report in "What We Know, and What We Have Not Yet Learned: Triple Disasters and the Fukushima Nuclear Fiasco in Japan," when the earthquake and subsequent tsunami knocked down the nuclear power plants owned and operated by the TEPCO (Tokyo Electric Power Company), the company's initial reports avoided using the term meltdown: "It was not until much later that the firm publicly admitted that the meltdown of radioactive rods had occurred just a few hours after the tsunami knocked out the Fukushima plant. During this time, the central government, including the chief cabinet secretary, only echoed information issued by the power company. The regulatory agencies failed to collect

independent data relative to the breakdown of the reactors in Fukushima." The Japanese government did not have a smooth, cooperative relationship with the nations who offered international help. In "Deconstructing the Zero-Risk Mindset: The Lessons and Future Responsibilities for a Post-Fukushima Nuclear Japan," Tatsujiro Suzuki argues, "Immediately after the earthquake and tsunami hit Japan, the international community responded, offering disaster relief, nuclear expertise, and nuclear reactor equipment, among other things. The problem was that Japan could not appropriately respond, because it lacked a specific office to accommodate such assistance. Communication with the international community was not always sufficient, as was evident when discharging water with low-level radio-activity into the sea without promptly and adequately informing neighboring countries." It seems that the new idea might be need to improve the people's ability to have temporal cooperative conditions with the other nations when they need. See Akira Nakamura et al., "What We Know, and What We Have Not Yet Learned: Triple Disasters and the Fukushima Nuclear Fiasco in Japan," *Public Administration Review* 71, no. 6 (2011): 897; Tatsujiro Suzuki, "Deconstructing the Zero-Risk Mindset: The Lessons and Future Responsibilities for a Post-Fukushima Nuclear Japan," *Bulletin of the Atomic Scientists* 67, no. 5 (2011): 14–15.

30. Paul Gilster, *Digital Literacy* (New York: Wiley, 1997), 154.

31. André C. Drainville, *Contesting Globalization: Space and Place in the World Economy* (London: Routledge, 2004), 7.

32. Hirschkop, *Mikhail Bakhtin*, 261.

33. R. Barri Flowers, *Domestic Crimes, Family Violence and Child Abuse: A Study of Contemporary American Society* (Jefferson: McFarland, 2000), 3.

34. In "Pupil Insights into Bullying, and Coping with Bullying" (2002), Tomoyuki Kanetsuna and Peter K. Smith inform: "Starting in Scandinavia, research on bullying and its related issues has now been conducted extensively in Europe, North America, Asia and Australia. . . . An interesting case is the comparison between 'bullying' in western countries, and 'ijime' in Japan. 'Ijime' is the Japanese term considered most similar to bullying, and just as there have been suicides in England caused by persistent bullying, so in Japan suicides due to 'ijime' have led to widespread social concern." See Tomoyuki Kanetsuna et al., "Pupil Insights into Bullying, and Coping with Bullying: A Bi-National Study in Japan and England," *Journal of School Violence* 1, no. 3 (2002): 6–7.

35. For example, in "Domestic Violence and Women's Mental Health in Japan" (2001), Rita Weingourt and other scholars argue: "In contrast to other Asian nations, Japan has the second largest economy in the world and a 90 percent literacy rate. Japan ratified the International Bill of Human rights in 1979 and the Convention on the Elimination of All Forms of Discrimination against Women in 1985. Despite great progress toward formal acknowledgement of the inequality of women, Japan is still lagging behind in identifying the scope of the problem and the health hazards caused by violence against women." In "Fatal Child Abuse in Japan" (2011), Saori Nambu and other scholars report that "it has been pointed out in Japan that criminal punishment in domestic homicide cases, especially in fatal child abuse cases, tends to be more lenient than in public homicide cases that occur outside the home." See Rita Weingourt et al., "Domestic Violence and Women's Mental Health in Japan,"

International Nursing Review 48, no. 2 (2001): 102; Saori Nambu et al., "Fatal Child Abuse in Japan: Does a Trend Exist toward Tougher Sentencing?" *Journal of Injury and Violence Research* 3, no. 2 (2011): 74.

36. Yi-Fu Tuan, "Introduction: Cosmos versus Hearth," in *Textures of Place: Exploring Humanist Geographies*, ed. Paul C. Adams, Steven Hoelscher, and Karen E. Till (Minneapolis: University of Minnesota Press, 2001), 320.

37. Hirschkop, *Mikhail Bakhtin*, 5.

38. Jürgen Habermas, *The Theory of Communicative Action*, vol. 1, *Reason and the Rationalization of Society* (Boston: Beacon Press, 1984), 50.

39. Hirschkop, *Mikhail Bakhtin*, 244.

40. Gergen, *The Saturated Self*, 227–28.

41. Faulkner, *Faulkner in the University*, 139.

42. Ibid.

43. Ibid.

44. Paul Giles, *The Global Remapping of American Literature* (Princeton: Princeton University Press, 2011), 214.

45. Thomas L. Friedman, *The World Is Flat: A Brief History of the Twenty-first Century* (New York: Picador, 2007), 5.

46. Jon Smith et al., "Introduction: Uncanny Hybridities," in *Look Away!: The U.S. South in New World Studies* (Durham: Duke University Press, 2004), 6.

47. John T. Matthews, "The Southern Renaissance and the Faulknerian South," in Monteith, *The Cambridge Companion*, 117.

Chapter 1

Progress from Monologism to Dialogism

Studying Two Fa(u)lkners' Works

Come, ye tyrants and haughty kings,
Stoop and listen here to little things;
I seize my pen to write a little story
Of *love*, *war*, *blood*, lust, and glory.

—W.C. Falkner, *The Siege of Monterey*[1]

The whole world breathes and calls to me
Who marble-bound must ever be.

—William Faulkner, *The Marble Faun*[2]

When an author is admired as a genius, it means that the author becomes a guardian of the particular realm in which his/her ideological function establishes cultural standards. In "What Is an Author?," Foucault explains that the "author function" works when an author's name represents certain kinds of cultural values that give particular criteria, limitations, or principles, arguing that "the author's name manifests the appearance of a certain discursive set and indicates the status of this discourse within a society and a culture." Idolizing an author enables readers to limit, choose, or exclude some particular cultural products as the result of their criticism. An author becomes a genius when the author becomes "a certain functional principle," or "an ideological product," which restricts and controls people's verbal activities and evaluations.[3] The author indicates a criterion with which the readers can tell what kind of discourses should be mocked or eliminated from their age.

Reading W.C. Falkner's works is one of the best ways to understand what kind of elements Faulkner excludes from his works to be a genuine "author" of his modern age. In the middle of the twentieth century, Faulkner is approved as the American South's cultural guardian, who denounces the late nineteenth and early twentieth century's romantic terrain and makes his great-grandfather's fiction a typical target of criticism by his contemporary culture. If Foucault's argument is correct, to understand Faulkner's author function, today's readers should know W.C. Falkner's works as the ideological counterpart to Faulkner's works—or, readers should know how, Faulkner cripples his great-grandfather's bumptious literary imagination following "the principle of thrift in the proliferation of meaning."[4] Faulkner's exclusion of W.C. Falkner's contexts and plots appears in the symbolic gesture of his name. Faulkner put a "u" in his original family name, William Cuthbert Falkner, when he published his first poem. Later, Faulkner explained in his private message to Malcolm Cowley that he added this letter because he "secretly was ambitious and did not want to ride on [great-]grandfather's coat-tails" (SL 212). Faulkner wanted to minimize the value of W.C. Falkner's literary legacy in his name and his works, recognizing that his great-grandfather's name was still big enough to influence readers' judgment on his writing. Juxtaposing W.C. Falkner and Faulkner as two bestselling authors in Mississippi offers us a chance to reconsider why today's readers, including ESL students, choose one of these authors as the foundation of Southern culture and not the other.

It is difficult, even for today's American readers, to rightfully retrieve the contexts of W.C. Falkner's romantic literary works within Faulkner's texts. As Richard Godden confesses in the beginning of *Fictions of Labor* (1997), reading Faulkner is difficult for US scholars because his fiction suppresses a tale that should be recovered by careful reading: "Faulkner's 'difficulty' . . . is driven by his penurious habit of secretion—a habit which demands that the reader attend closely in order to recover, from Faulkner's choked, subverted, underarticulated, and yet imperious prose, inferences of a tale that is not being told."[5] Godden's argument suggests some secret "tale" can be found behind Faulkner's laborious rhetoric; however, if readers analyze the historical context in Faulkner's works, it is possible to assume that the "tale" Faulkner suppressed behind his difficult prose is the conventional pattern of late nineteenth-century Southern romance, which is typically represented by W.C. Falkner's works. Faulkner shrugs off this "tale" not because of his "penurious" habit, but because the conventional plots of Southern romance are too obvious for him, and because his great-grandfather's works are too familiar and major for him.

Or maybe it is possible to say that Faulkner deals with the Southern narrative's self-contented heroism by reading W.C. Falkner as if he were naive.[6] The Southern heroism requires the hero to renounce his present life, in order to imagine his heroic death in the battlefield or to attain his ideal life in his romantic writing. The Southern heroes—which include W.C. Falkner who took part in the Civil War, his romantic male characters, young Faulkner who tried to go to the real battlefield in the World War I, and the self-destructive protagonists in his early novels—look upon themselves as dead men in their imaginations, to proudly overview their whole lives in their complete forms. This presupposition of the heroic death is a typical pattern of the Southern heroes' self-satisfied lives. They have already imagined their whole lives before they begin them, and then they launch their lives to fulfill them. However, Faulkner, who survived World War I and was sick and tired of this self-poisoning obsession, objectified his local area's popular romanticism and recaptured the people's modern lives in the current flow of time. Faulkner may have not completely relinquished his propensity for hero worship, but to escape from this innate tendency, he intentionally treats the Southern heroism as a naive one. Therefore, even when what today's readers want to know is not Foucauldian historicized text that threateningly proliferate Faulkner's works, but what kind of gravity—what kind of Derridean difference—is working "under erasure [*sous rature*]" of Faulkner's texts, reading W.C. Falkner's works is still beneficial for learning what kind of "literary suicide" happens, what kind of "desire of presence" takes place, and what kind of "non-fulfillment" drives the frustrated characters' impulsions in Faulkner's texts.[7] Faulkner, as a modern author, disdained W.C. Falkner's minor texts, but it scarcely denies that Faulkner was saturated in the self-indulgent mode of the fantastic Southerner's romantic symbolism.

Juxtaposing two Fa(u)lkners' works was not such an eccentric idea for Mississippi's local readers in the last century. In fact, in mid-twentieth-century Mississippi, both W.C. Falkner and William Faulkner are counted as bestselling authors. In "Book Reading in the Senior Years: The Habits and Preferences of 200 Mississippians" (1960), Jere Hoar interviews 200 seniors in Mississippi who do not read much, and reports that the eleven books mentioned by more than one respondent as having been the last book they read include W.C. Falkner's *The White Rose of Memphis* and Faulkner's *Intruder in the Dust*.[8] In 1960, for the ordinary seniors in Mississippi, the works of both W.C. Falkner and Faulkner are included in the few bestsellers they read. For example, Hoar mentions an 82-year-old woman who "last read *The White Rose of Memphis* 40 years ago to her children." *The White Rose of Memphis* is remembered by Mississippi's general audience as "a best seller written some 75 years ago by Col. W. C. Falkner, grandfather of the present day Nobel Prize winner who lives in the community."[9] The twentieth-century community in Mississippi still listed W.C. Falkner's romantic novel in the same vein with Faulkner's works.

For ESL students, it is beneficial to juxtapose W.C. Falkner's works and Faulkner's works in order to analyze their historical development. These novels are not the foundation of the American South but are a part of a competitive historical process to recompose the cultural structure of the area. In what way can we imagine the American South? Is it a land dominated by imperialistic heroes, or is it a land of personal connections between diversified individuals? Resisting W.C. Falkner's romantic vision of the American South, in *Intruder in the Dust*, Faulkner uses the socially marginalized characters' personal combinations to promote the community's autonomous potential to create egalitarian relationships that resist the monolithic white male community. In this case, the young white protagonist's desperate resistance against the community's white supremacy is difficult to understand for ESL readers if they do not know how the protagonist and the other white male characters in Faulkner's story are sickly contaminated by the Old South's romantic imagination in the middle of the twentieth century. ESL students should understand why Faulkner's white male adults "cant listen" (ID 70) to the socially marginalized people's words, and why they can overlook the segregated condition of African Americans. If Yoknapatawpha's dominant group is the white male adults who share the same imagination, what kind of imagined world do they share to create their monolithic community? An effective way to answer this question might be to read the works of Faulkner's great-grandfather.

One of the most frequently referenced literary figures in the studies of Faulkner's works is his great-grandfather, W.C. Falkner, a popular author of several romantic works as well as a lawyer, soldier, and businessman. However, this does not mean a full comparison of these two author's works has been done by Faulkner scholars. H.L. Mencken's iconoclastic essay, "The Sahara of the Bozart" (1917), argues that the South from the late nineteenth century to the early twentieth century is a cultural wasteland. Mencken's influential words have allowed modern Southern writers to censure and ignore their literature's romantic legacies. Faulkner, as a modernist author in the twentieth century, scorned the sentimental works in the Old South, and his criticism's main target was his great-grandfather's works. In "The Faulkners: Recollections of a Gifted Family" (1963), Robert Cantwell reports that when he had an interview with Faulkner in 1938, Faulkner was glad to talk about "Colonel" W.C. Falkner, but outside of his ancestor's military career he seemed to have little interest; Faulkner thought his great-grandfather's *The White Rose of Memphis* "was pure escapism" and "the Colonel had no humor and probably no sensibility."[10] Faulkner's cold attitude toward his great-grandfather's fiction reflects modern scholars' critical reactions to the Southern romances of the late nineteenth to the early twentieth century. Faulkner and the twentieth century's main critics have successfully disdained W.C. Falkner as a common, sentimental author and have failed to realize that W.C. Falkner's works are one of the fictional roots of his great-grandson's writing. They have agreed to disregard W.C. Falkner's works and celebrate Faulkner as the sudden appearance of sheer genius from nothing.

The history of criticism has followed this tendency to enforce the myth of Faulkner's isolated talent. When Frederick J. Hoffman and Olga W. Vickery complete the 1960 edition of an authorized anthology of Faulkner studies, *William Faulkner: Three Decades of Criticism*, Hoffman recollects that—from when Faulkner's reputation began in 1939 and rapidly accelerated with Cowley's editing of *The Portable Faulkner* in 1946—Faulkner is rediscovered as "the obscure and isolated genius"[11] who symbolizes the South's cultural revival, so-called the Southern Renaissance,[12] to fight the negative image that the South cannot produce new talent. This discovery is connected with the idea of independence in Faulkner's literary world. As Mark Royden Winchell argues in *Cleanth Brooks and the Rise of Modern Criticism* (1996), since Malcolm Cowley propelled Faulkner's reputation with *The Portable Faulkner* in 1947,[13] one of the main trends in Faulkner study included seeing his fictive world as a complete universe, "the Yoknapatawpha saga whole."[14] Cleanth Brooks's *William Faulkner: The Yoknapatawpha Country* (1963) is the first book, which critically reads Faulkner's literary world as an independent universe that "embodies its own principles of order."[15] In these studies, Faulkner's novels are disconnected from his great-grandfather's second-rate literature. Faulkner is portrayed as a solitary talent who created his "mythical kingdom" by himself.[16] But is it really true that a genius can appear from nothing? Is he an original creator of his own literary world?

Though many scholars neglected the relationship between W.C. Falkner's works and Faulkner's, one scholar began his important study in the 1960s. Duclos's *Son of Sorrow* is still now considered one of the finest studies of W.C. Falkner's life and literary works. When Duclos wrote "the most thoroughly researched biography"[17] of W.C. Falkner as his dissertation in 1961, he was one of the few critics who actually had contact with Faulkner and recorded his family's testimonies about W.C. Falkner's life and works.[18] In his investigation of W.C. Falkner's life and writing, Duclos points out common motifs of both Fa(u)lkners' works and finds "striking parallels and similarities in the works of both men." Duclos proves that some motifs show interesting similarities between these two authors' works; however, he admits his study does not clearly show in what way Faulkner shares and resists W.C. Falkner's romantic motifs. He encourages future studies of two Fa(u)lkners' works, arguing "only a more disinterested observer can detect the similarities, which go further than their common name or even more than this study has been able or can reveal."[19] Duclos finds some motifs of Faulkner's works show the author's critical struggles with his great-grandfather's literary world. W.C. Falkner's literary motifs are twisted and re-accentuated in Faulkner's works, but these relationships have not been fully explored so far. Duclos's work was formally published to the public in

1998. It might be today's critics who can establish appropriate methods to analyze the historical development using theoretical frameworks to critically compare both Fa(u)lkners' writing.

To analyze Faulkner's artistic struggle with his great-grandfather's literary works, this study mainly uses Bakhtin's theoretical terms. The first couple of terms are internally persuasive discourse and authoritative discourse. According to Bakhtin, an author struggles with historically authorized discourse to establish his own expressions. An author's competition with orthodoxy can be considered a struggle of "an *internally persuasive discourse*" against "*authoritative discourse*" (DN 342). If an author wants to consciously establish his own persuasive style, the author's expressions should be "interanimating relationships with" the world's current contexts to resist against authoritative discourse (DN 346). Bakhtin argues: "When thought begins to work in an independent, experimenting and discriminating way, what first occurs is a separation between internally persuasive discourse and authoritarian enforced discourse" (DN 345). In Faulkner's case, this authoritative discourse is represented by a romantic pattern in W.C. Falkner's works.

Authoritative discourse is not directly mentioned or represented in an author's internally persuasive discourses. As Bakhtin argues, as if it was "the word of the fathers," a particular community's conventional discourse is "located in a distanced zone, organically connected with a past that is felt to be hierarchically higher" (DN 342). Authoritative discourse is suggested in an author's iconoclastic mockeries or parodic profanations. This is an author's artistic re-accentuation of the authorized discourse, which establishes the author's intentions by re-narrating the authorized discourse in different ways. In this satirizing attitude, modern novels' discourses since *Don Quixote* are considered to be "several types and variants of double-voiced discourse" that have "a parodic quality" (DN 419). Especially in his early and middle works, Faulkner wrestles with W.C. Falkner's romantic worldview and creates his own compelling, modern worlds by parodying some of W.C. Falkner's motifs.

This chapter also uses Bakhtin's terms monologism and dialogism to explain the development from W.C. Falkner's romantic works to Faulkner's modern works. Monologism, in which the author's voice is identified with the narrator's and the hero's voices to totally dominate the other voices, is strongly contrasted with dialogism, in which various voices interrelate to make an art form that never denies social diversities. Bakhtin's two works, "Discourse in the Novel" and *Problems of Dostoevsky's Poetics* suggest that readers can employ monologism and dialogism to explain two types of literary forms. According to Bakhtin, monologism is defined as a consciousness that, "at its extreme, denies the existence outside itself of another consciousness with equal rights and equal responsibilities, another *I* with equal rights" (PDP

292). Monologism is an extreme form of narrative in which the narrator and the hero are identified with the author and continue their own self-affirming words. It is a narrative approach that strengthens a subject's ideally unified vision. Monologism appears in romantic poems and novels in which the author's worldview is directly reflected in the characters, and the characters' actions and utterances are straightway connected with the author's intention. Typical monologism shows no distance between the author's consciousness and the hero/narrator. This is "the typical path taken by a monologic novel of the romantic type" (PDP 12). In this worldview, a hero's utterance is identified with the author: "The hero is merely the carrier of an independently valid idea; as a true signifying idea it gravitates toward some impersonal, systemically monologic context; in other words, it gravitates toward the systemically monologic worldview of the author himself" (PDP 79). A romantic novel's author disregards the existence of another consciousness to keep its artistic consistency. In monologism, the author's idea finalizes the whole development of the plot.

In contrast to the romantic novel's monologism, Bakhtin studies Fyodor Dostoevsky's multivoiced novel as a typical literary model of dialogism. In dialogism, multiple voices in various social strata "enter the novel and organize themselves within it into a structured artistic system" (DN 300). Dostoevsky's fiction is dialogic—it "is constructed not as the whole of a single consciousness, absorbing other consciousnesses as objects into itself, but as a whole formed by the interaction of several consciousnesses, none of which entirely becomes an object for the other" (PDP 18). Characters have their own lives and voices that will never be unified in a happy ending: "In Dostoevsky's novels, the life experience of the characters and their discourse may be resolved as far as plot is concerned, but internally they remain incomplete and unresolved" (DN 349). In the last scene in Dostoevsky's *Crime and Punishment*, for example, the murder case is resolved, but Raskolnikov's moralistic struggle for redemption, or his "internal dialogism" (DN 350) continues. Because of this continuation of dialogic drama in every character's internal self, "great novelistic images continue to grow and develop even after the moment of their creation" (DN 422). In short, while monologism disregards the existence of another consciousness to maintain its literary world's consistency, dialogism celebrates a protagonist that continues internal debates (with several voices) even after the resolution of the plot. Bakhtin argues that the novel's style has developed from monologism to dialogism with the deepening and broadening of the world.

Pure monologism is an abstract idea that cannot be realized in our actual world because every word is always shared with other people's utterances to create social meaning. As Todorov argues, it is not really possible "to maintain a distinction between a discourse of dialogical nature and a monologic discourse, since every discourse is, by its very nature, 'dialogical,' that is, caught in intertextual relations."[20] However, monologism is a useful concept when criticizing a dominant voice's suppressive power over the other voices in a literary work. As Bakhtin argues, in the monologic worldview, the thoughts and ideas that are different from the author's point of view are not affirmed, and these different viewpoints are "either polemically repudiated, or else they lose their power" (PDP 80). In W.C. Falkner's works, the characters that stand against the heroes are labeled villains and foes, while the characters that support the hero are categorized as insiders in the hero's narrow community that realizes the author's ideal world.

W.C. Falkner's monologism finally excludes marginalized people's ability to have social combinations. W.C. Falkner's fiction inherits the literary tradition of the antebellum South's "unfinished fairy-story,"[21] in which the chivalrous heroes love the idealized heroines and make a hierarchical world that establishes the white male at the top of the pyramid. Each of W.C. Falkner's works obsessively preaches that the hero's unified world vision will finally be fulfilled. This monologic worldview positively affirms the hero's belief in violent passion, white supremacy, and patriarchy. Under the pressure of this arrogant idealism, the minor characters' temporally democratic relationships with the hero are regarded as small episodes in the process to achieve his ambitious goal. This vision also represents W.C. Falkner's desire to establish his own eternal glory when he dies, in the same way his own statue established beside his grave finalizes his whole life.[22] Both in his literary world and in real life, W.C. Falkner embodies the advantage of masculinity that will endure hardships to attain a successful life. A monotonous, white male-oriented worldview saturates the whole of W.C. Falkner's works to ignore undesired perspectives as worthless digressions. Indeed, as Thomas Felix Hickerson argues in *The Falkner Feuds* (1964), W.C. Falkner's personality is "arrogant (but intriguing)."[23] This arrogant yet intriguing psyche is the origin of W.C. Falkner's success, as well as the main curse of his great-grandson's fiction.

W.C. Falkner's egoism is attractively embellished in his fiction. In each of his literary works, the author identifies his worldview with a hero who obtains wealthy social status and dominates his narrow community with his self-centered attitude. This imagined heroic image is an archetype for the author's successful self-image in his community. This self-centered monologism's mythic effect that governs W.C. Falkner's life and writing is one of the main targets of his great-grandson's criticism. Why does the illusion of a masculine hero still survive in the modern world? Is it true that a white male's iron will promise a future prosperity that will be eternally inherited by his children's children? Is there any righteousness in the white male dominance of their narrow communities? These questions influence the obsessive themes of Faulkner's early and middle works, from *Soldier's Pay* to *Go Down, Moses*. In other words, Faulkner's early and middle works continually have a dialogic relationship with W.C. Falkner's monologic imagination that deeply influences the community's everyday life. As Quentin's case in *The Sound and the Fury* and *Absalom, Absalom!* typically demonstrates, some of Faulkner's characters who are obsessed with death are attracted to this monologic voice that uniquely represents the American South's past and glory.

A historical function of W.C. Falkner's monologism is to establish the double-voiced formation in Faulkner's fiction. Faulkner's main characters—young Sartoris Bayard in *Flags in the Dust*; Quentin Compson in *The Sound and the Fury* and *Absalom, Absalom!*; Joe Christmas, Joanna Burden, and Gail Hightower in *Light in August*; Drusilla Hawk Sartoris and Bayard Sartoris in *The Unvanquished*; and Ike McCaslin in *Go Down, Moses*—are obsessed with a romantic worldview that demands an eternal world under the dominance of the white male hero. Faulkner's modern characters ironically refrain and convert the old images of romantic heroes and heroines, who are prototypically shown in W.C. Falkner's novels. Using Bakhtin's term, Faulkner's dialogic relationship with his great-grandfather's texts can be called "double-voiced discourse" (PDP 192), in which Faulkner's text is a reaction against his great-grandfather's authoritative discourse. W.C. Falkner's words are indirectly introduced into Faulkner's texts as the authoritative discourse that forms new meaning. Faulkner's experimental literary forms do not have any stylistic stability, but he uses these forms for multiple examinations of the American South's historical heritage that is represented in W.C. Falkner's works. The similarity of the motifs of both author's texts tell readers how Faulkner critically inherited the romantic pattern in W.C. Falkner's writing and examined them under new lights.

In this historical development from W.C. Falkner's fiction to Faulkner's writing, characters' social combinations appear in both authors' texts. In W.C. Falkner's romantic works—from his epic poem, *Siege of Monterey*, to his last novel, *The Little Brick Church*—some characters show their ability to make social combinations in the processes of the plots, but all these possibilities are eventually demolished under white male dominances. In each work, all the relationships are ordered from top to bottom in a pyramidal hierarchy, and in the end the unified couples and families are idolized. In every W.C. Falkner story, the top is occupied by the ideal white couple, and the other characters are categorized as friends, supporters, subordinates, and the admiring narrator, while the antagonists are summed up with the words "foe" and "villain." Especially in *The Little Brick Church*, the hero is established as an independent American that protects and dominates the weak and the helpless. In this romantic tale, the hero's ideal goal is his complete unity with his heroine and the establishment of his dominance in his narrow community in order to protect that community from foes and villains. The hero and heroine's tragic death on the battlefield of the American Revolution is an ideal death for them to attain this goal, because the hero has already gained the heart of the heroine and they show their pure will to protect their communities from the enemy in battle. Moreover, as the narrator's favorable attitude toward the plantation culture of the American South suggests, the Southern white males are allowed to believe they properly inherit the American liberal spirit to protect the weak and helpless. This paternal attitude excuses their arrogant treatment of women, children, and African Americans. W.C. Falkner's works directly express the logic of the romantic white male's imperial attitude in the American South.

Critics disregarded W.C. Falkner's works in part because of the inaccessibility of paper copies of his works. Most of W.C. Falkner's works were not well circulated during the twentieth century. However, this condition is improved today because of the digitization of nineteenth-century texts. Readers today can find some of W.C. Falkner's works on web archives. The Library of Congress's website says W.C. Falkner's first published poem, *The Siege of Monterey* (1851), is "also available in digital form on the Internet Archive Web site."[24] The full PDF file of W.C. Falkner's first novel, *The Spanish Heroine* (1851), is offered on Google Books.[25] The New York Public Library tells a PDF of *The White Rose of Memphis* (1881) is found on the HathiTrust Digital Library—its contents include digitized texts from Google Books, the Internet Archive, and local libraries.[26] And the Library of Congress indicates the HathiTrust Digital Library for a PDF copy of Falkner's travel sketches, *Rapid Ramblings in Europe* (1884).[27] Some of his works are not available on the web. His drama script, *The Lost Diamond* (1867), cannot be found on the web—it is also impossible to find a paper edition. The world's largest bibliographic database, WorldCat, has no results when this book's title and author is searched. According to Duclos, "It is unfortunate that a text of the play [*The Lost Diamond*] has not been found. . . . Perhaps a copy of the play may never be found."[28] The other one, the full text of Falkner's last novel, *The Little Brick Church* (1882), cannot be found on the web, mainly because it is still available in hard copy.[29] In today's digital environment, especially for international students, it is regrettable that *The Little Brick Church* has not been digitized on the web, because W.C. Falkner most strongly illustrates his philosophy and worldview in this last romance. However, the paper edition, which was reissued from Literature House and Gregg Press in 1970, is well circulated and easy to find in local libraries. Almost all of W.C. Falkner's works are accessible for today's ordinary readers who can use the digital network and its information.

On the other hand, Faulkner scholars make huge efforts to digitally archive the detailed information about Faulkner's fiction in some Internet sites. The University of Mississippi's *William Faulkner on the Web* offers rich information about Faulkner's works, biographical studies, geographies, films, and useful details.[30] *Digital Yoknapatawpha*, which is made by a collaboration of Faulkner scholars with the Institute for Advanced Technology in the Humanities, the Digital Media Lab and SHANTI at the University of Virginia, offers visually geographic explanations on Faulkner's works. Its "Visualizing Character and Location in Faulkner" page offers a set of visualizations that illustrate, from a computational perspective, the relationship between people and places in each of William Faulkner's Yoknapatawpha fictions.[31] R.P. Stoicheff's hypertext version of William Faulkner's *The Sound and the Fury* is now deleted from its homepage but it was a classic example that translated Faulkner's modern "proto-hypertext" work into "a freely accessible hypertext edition on the web" with useful links to intertexts, criticisms, and visual displays.[32] The full audio records of Faulkner's sessions with audiences at the University of Virginia in 1957 and 1958 are excellently archived in the University of Virginia's site, *Faulkner at Virginia: An Audio Archive* and people can freely access this site and hear Faulkner's voice.[33] Thanks to these sites, various information pieces about Faulkner's works are easily accessible to current readers; however, these fertile digital resources make little effort to connect W.C. Falkner's works and Faulkner's works. In the near future, Faulkner's whole literary works will belong to the public domain and be digitally archived. Both W.C. Falkner's and Faulkner's works will be juxtaposed in the digital space, and the way to connect them again will involve reconsidering their literary development in the history of the American South.

This chapter argues the relationships between these two author's works are the process of Faulkner's struggle to establish his own literary world against W.C. Falkner's authoritative monologism. In the first part, I analyze how several of W.C. Falkner's motifs are insinuated, transformed, and criticized in Faulkner's works. In the second part, I examine how W.C. Falkner's biographical records are adorned by his contemporary biographer and how that discourse is amplified with his fiction. In the third part, I compare W.C. Falkner's first poem, *The Siege of Monterey*, with Faulkner's *Absalom, Absalom!* to prove how Sutpen's design is a re-accentuation of the romantic protagonist's life in W.C. Falkner's epic poem, as well as how modern narrators, Quentin Compson and Shreve McCannon, are related to Sutpen's monologism. In the fourth part, I compare Faulkner's short story "Wash" with W.C. Falkner's short novel, *The Spanish Heroine*, to prove the race and class conscious is mentioned in both works. In the fifth part, I briefly argue the divine image of the decent families in W.C. Falkner's bestseller novel, *The White Rose of Memphis*, and Faulkner's *Flags in the Dust*. In the sixth part, I compare W.C. Falkner's *The Little Brick Church* with Falkner's *Intruder in the Dust*. This part explains how the white hero in *The Little Brick Church* considers his own dominant power to protect the other weak and poor as his liberal spirit, and argues that the modern Southern elite lawyer Gavin Stevens inherits this Southern white male hero's typical attitude while the young protagonist Chick Mallison reconfigures the American South as a place where individuals can form egalitarian relationships and keep their community's changeability. This entire chapter examines how Faulkner's texts keep their dialogic relationship with the authoritative voices that are represented in W.C. Falkner's works.

FROM W.C. FALKNER'S MONOLOGISM TO FAULKNER'S DIALOGISM

All of W.C. Falkner's literary works share the same romantic pattern: loving white couples separate from each other for a while, but reunite in the end. What the hero and the heroine want is to stay together from their relationship's beginning to end and to be together indefinitely in heaven. In many cases this pattern is concluded by their marriage at the end of the story, suggesting their eternal unity. The obstacles that separate their unity are overcome by their violent struggles. Other various elements in the stories—nations, races, classes, wars, technologies, travels, and gender disguises—are arranged to give dramatic variety to this main plot in each work.

The locations and historical ranges of W.C. Falkner's works are broad, but his worldviews are quite narrow because of this monotonous pattern of the main plots. Readers can enjoy the various ethnic stages of W.C. Falkner's works. In *The Siege of Monterey* and *The Spanish Heroine*, heroes and heroines both take part in the US-Mexican War. The location of *The White Rose of Memphis* is a nineteenth-century masquerade ball on a luxurious steamboat in the Mississippi. In *The Little Brick Church*, a centenarian narrates his memory of a couple whose love ends in tragic death on the battlefield during the American Revolutionary War. W.C. Falkner himself travels around Europe and reports various European landmarks in *Rapid Ramblings in Europe*. W.C. Falkner permeates various times and places with his monotonous romantic imagination. All his works end with the eternal bliss of loving couples, suggesting that all obstacles are conquered to realize one ideal unity.

These romantic relationships are hazardous for today's readers because they prevent the realization of the characters' egalitarian relationships. W.C. Falkner's plots never aim to realize the equal relationship between characters. Rather, his romanticism works to naturally reproduce the white male dominance in his narrow community. For example, in his last novel, *The Little Brick Church*, W.C. Falkner describes the heroine's total belonging to the hero's vision. Oscar, the hero—who is the son of a landowner family in Colonial America—and Olivia, the heroine—who is a descendant of an English aristocrat—suggest an egalitarian relationship in the new world, but Oscar's words absorb Olivia's voice when they decide to marry. Before he goes to fight in the American Revolutionary War, Oscar asks Olivia to marry him after the battle, on Christmas day. Their conversation demonstrates what kind of monologic conversation is idealized in W.C. Falkner's romantic discourse:

> When the hour of separation arrived, Oscar drew her [Olivia] fondly to his heart and endeavored to dispel the dark, gloomy cloud of sorrow that lowered over her mind. Smiling through her tears, she said,—
>
> "If God will hear my prayers He will protect you from the enemy's bullets, for I mean to constantly besiege His throne in your behalf."
>
> "Yes, darling," replied Oscar, "and if He and you are willing, we shall be married on Christmas-day. What do you say to it?"
>
> Nestling her pretty face close up under his chin, while her beautiful cheeks grew crimson with a modest blush, she, in a whisper, replied,—
>
> "I have no will, no mind, nor heart of my own; they all belong to my Oscar; use them as you please." (LBC 392–93)

W.C. Falkner confirms that the couple's ideal goal is the heroine's complete unification with hero's imperial worldview. Oscar's demand for complete obedience from Olivia is naturalized without hesitation, and the heroine obeys the hero's will completely. Olivia's devotion of her life to Oscar finally causes her tragic death. At the end of this story, Olivia accidentally falls in and drowns in a frozen pond trying to save Oscar from the death penalty on the battlefield. Oscar escapes the execution, but after discovering his fiancée's death he has a heart attack and dies beside the body of Olivia. This ridiculous, but tragic ending of a young couple is idealized in this novel.

This romantic conversation of hero and heroine in *The Little Brick Church* shows the typical romantic pattern that influences the white male Southerners' dominance over women in Faulkner's works. For example, Faulkner depicts the white male Southerner's aspiration for romantic dominance of women in *The Sound and the Fury*. In his incestuous desire for his sister, Caddy, Quentin demands complete obedience from her, wishing to keep his family's unity. Caddy ignores Quentin's demand and sleeps with a veteran soldier instead, Dalton Ames. The conversation between Quentin and Caddy after Quentin's discovery of Caddy's unchasteness shows the double-voiced effect of W.C. Falkner's romantic discourse and Faulkner's parodic modern voice. When Quentin orders Caddy, "go on to the house like I told you," Caddy sarcastically repeats a conventional response of the obedient woman to soothe Quentin's irritation: "yes Ill do anything you want me to anything yes" (SF 997). Caddy mocks the submissive heroine to acknowledge Quentin's romantic illusion for a while; however, soon after this conversation, Caddy resumes her desire for the veteran soldier and says, "Im bad anyway you cant help it" (SF 998). Caddy's furious resistance against Quentin also shows her will to resist the authoritative discourse in Yoknapatawpha, which Quentin feels is "a curse on us" (SF 998). Quentin recognizes that Caddy is not an obedient heroine but a modern individual figure that disturbs the Southern conventional mores, but he cannot help imposing the romanticized heroine's image upon Caddy. Caddy's voice is not only responding to Quentin's voices, but is also re-accentuating and resisting the authoritative voice of the antebellum South's imagined world.

In W.C. Falkner's works, the intimate relationship between childhood sweethearts is a kind of pseudo-incestuous relationship. In *The Little Brick Church*, Oscar knows that Olivia is adopted, but they are brought up as brother and sister. Oscar calls Olivia his "pretty sister" (LBC 141), and when he is informed that the Duke of Moreland, Olivia's father in England, has tricked her into a marriage with Lord Nordane, Oscar furiously accuses Nordane's moral decadence and Olivia's unchasteness. Considering himself and Olivia to be inseparable, Oscar insists: "We were nursed at the same breast, rocked in the same cradle, sheltered by the same roof, baptized in the same church, taught at the same school, swam in the same creek, learned to lisp our first prayer while kneeling by my mother's side, and our souls had but one thought, and that was of each other" (LBC 341). As a result of this strong connection, Oscar demands that Olivia be his wife. Oscar's strong love for Olivia is quite similar to Quentin's incestuous obsession with Caddy.

The lovers' nearly incestuous relationship in W.C. Falkner's romance is naturalized as a hero's romantic desire to completely identify the relationship between himself and his heroine. This romantic relationship, however, is re-accentuated and parodied with Quentin's false claim of an incestuous relationship with his sister, revealing such romantic desires' unnaturalness. Quentin says, "I have committed incest" (SF 936) to his father when he discovers Caddy's physical relationship with Dalton Ames. In reality, Quentin's romantic love for his sister is severely rejected by Caddy's resistance against Quentin's caress with her harsh words, "*Dont touch me*" (SF 965). Quentin still demands to keep his sister under his control forever, and imagines Caddy and himself in hell because of their sinful relationship. In his illusion, Quentin hopes that in hell Caddy "*will have only me then only me then the two of us amid the pointing and the horror beyond the clean flame*" (SF 966). This imagination of hell is an apt penalty for Quentin—it is a reversed version of the romantic lovers' eternal marriage, which is depicted in *The Little Brick Church* as the tragic couple's eternal connection in heaven. The tragic end of Oscar and Olivia is celebrated by the preacher as their wedding day: "God has prepared for them a marriage feast in heaven, and their honeymoon shall last through eternity" (LBC 422). In W.C. Falkner's romantic imagination, God and the preacher both celebrate the lovers' tragic death and their unity, naturalizing Olivia's devotion to her lover. Quentin's desire is an ironic reversal of this romantically eternalized image.

The illusion of agreement between the man and woman through violence is also the target of Faulkner's criticism in his early and middle works. At the last scene of *Mosquitoes*, Faulkner parodies the heroine's total agreement with the hero's exercise of his imperial will through violence. In the end of this novel, a thirty-eight-year-old bachelor, Talliaferro reveals his desire that demands people willingly agree with his self-centered world. Despairing in his communications with young artists and his love, Talliaferro is aware of his violent desire to dominate them all. He feels he has been "too gentle" with them and his woman; he claims the most effective technique to control them is the "oldest technique in the world: a club" (MOS 540). He telephones his male friend, Fairchild, and tells him he has found a way to solve the whole problem, but he hears an unknown woman's voice that affirms his dangerous desire:

> [Talliaferro] "I learned through a mistake tonight. The trouble is, I haven't been bold enough with them: I have been afraid of frightening them away. Listen: I will bring her here, I will not take No for an answer; I will be cruel and hard, brutal, if necessary, until she begs for my love. What do you think of that? . . . Hello! Fairchild? . . ."
>
> An interval filled with a remote buzzing. Then a female voice said:
>
> "You tell 'em, big boy; treat 'em rough." (MOS 540)

Here Faulkner is mocking the typical romantic scene of nineteenth-century Southern novels, which boldly shows the heroine's complete agreement with the hero's violent passion. In W.C. Falkner's *The Spanish Heroine*, Ellen views her lover, Henry, "mounted on his white steed, a reeking sword in his hand, dealing death and destruction into the bleeding ranks of his country's foes" as "her pleasant dream" (SH 85). Talliaferro hears a woman's voice that encourages his brutal will to govern the people and a woman, but its unfamiliar voice implies that underneath Talliaferro's sudden rage lurks the old romantic notion of love.

In every W.C. Falkner's work, all the battles, miseries, and struggles make sense when the hero and heroine attain their eternal unity and prosperity. As Van Wyck Brooks suggests, "In the South much of the reality of the past was utterly forgotten or swallowed up and blurred and confused by legend."[34] This means that in the romantic imagination of the American South, the past is recomposed to promise the hero's unity with his heroine and future prosperity. Faulkner's modern novels parody W.C. Falkner's romantic motif to reveal that the literary tradition of the American South disregards the historical struggle after the Civil War. Faulkner's early and middle works reveal how the white protagonists reproduce its romantic myth in their obsessive voices and distort their society's practical development.

In Faulkner's works, his great-grandfather's romantic vision is suggested in the obsessive imagination that governs his modern protagonists. And this also means that W.C. Falkner's narcissistic vision is criticized from the other socially marginalized members' critical views in Faulkner's modern age. This dialogic condition between plural viewpoints re-accentuates his great-grandfather's old motif. As Bakhtin argues, a motif changes its meaning in different eras' dialogic conditions: "In an era when the dialogue of language has experienced great change, the language of an image begins to sound in a different way, or is bathed in a different light, or is perceived against a different dialogizing background" (DN 420). In the same way medieval chivalry stories are ironically reactivated in Cervantes's works, W.C. Falkner's romantic vision is insinuated, re-accentuated, and subverted in Faulkner's modern dialogic conditions.

In W.C. Falkner's mythic vision, the real records of the author and his romanticized images are interfused. W.C. Falkner intentionally connects his life with his literary works to embellish his real life with romantic visions. When Faulkner is "glad to talk about his great-grandfather," it means he likes to talk about his great-grandfather's life that is decorated with romantic imaginings. Faulkner thinks W.C. Falkner is "an overbearing man" and he recreates what W.C. Falkner actually aimed for in his modern characters' obsessions. [35] The next section briefly examines W.C. Falkner's life to illustrate how he combined his own life story with his romantic tales.

W.C. FALKNER'S LIFE AND HIS BIOGRAPHY

W.C. Falkner's romantic imagination had a powerful influence on his life. His biography was narrated as a heroic person's life, and Faulkner heard a romanticized tale of his great-grandfather from the country people and his relatives when he was growing up. This section explores how W.C. Falkner was portrayed as a heroic figure in his community's discourse in the beginning of twentieth century, and how this view has been corrected in later biological studies.

The biographical studies of W.C. Falkner that today's readers can access include these five works: Alexander L. Bondurant's "William C. Falkner, Novelist" (1900); Robert Cantwell's "Introduction" for the reprinting of *The White Rose of Memphis* in 1953; Donald Philip Duclos's *Son of Sorrow: The Life, Works and Influence of Colonel William C. Falkner, 1825–1889* (1998);

Thomas Felix Hickerson's *The Falkner Feuds* (1964); and Joel Williamson's *William Faulkner and Southern History* (1993). These works agree to tell W.C. Falkner's ambitious attitudes in the areas of education, law, military, business, politics, and literature, but they vary in each career's credibility.

Some of Bondurant's information sources in "William C. Falkner, Novelist" are W.C. Falkner's descendants, relatives, and friends.[36] Their image of "Colonel" W.C. Falkner is of a chivalrous confederate soldier, capable lawyer, and "business man"[37] who successfully built the Ripley Railroad, which extends from Middleton, Tennessee, to Ripley, Mississippi. Most of these idolatrous images are corrected in later biographies, though Bondurant's work illustrates how W.C. Falkner's life is officially recorded as "a representative mirror of the old South."[38] What William Faulkner heard in his childhood must have been this officially propagated image of his great-grandfather.

Bondurant's short biography is a column in *Publications of the Mississippi Historical Society* (1900). This column describes W.C. Falkner as a heroic military man, lawyer, entrepreneur, and novelist. Its biographical information is full of errors and idolizations, but worthy to summarize here because it interestingly shows how W.C. Falkner's legendary life is locally narrated in 1900.

William C. Falkner was born in Knox County, Tennessee, on July 6, 1826. From Tennessee his family removed to Missouri, and there his father died. When he was seventeen, he abandoned Missouri to go to Ripley, Mississippi—which Bondurant calls a pioneer, inland town in "a virgin land"—to meet his relatives. His relatives did not do much to raise the young man, and his self-education was personally helped by a teacher in the village. Because of his diligence, he gradually increased his social position in the community. Teaching elementary school while learning law by himself, he passed his examination to be a lawyer. He married Holland Pearce in 1847, but she survived only a few years, leaving one child, J.W.T. Falkner, who later became a member of the state senate and Faulkner's grandfather. In the US-Mexican War (1846–48), W.C. Falkner volunteered as a private, eventually being promoted to officer. Upon his return from the war field, he engaged in planting the land. He became "a slave owner, but was always a humane master." He remarried Elizabeth Houston Vance, of Alabama, in 1851. He became a member of the local Whigs, *The Know Nothing* party, which insisted that "America should be for Americans, and that in order to have a homogeneous Anglo-Saxon community the waves of foreign immigration that threatened to engulf the State, must be arrested." He strongly believed in the American South's right to secede, and during the Civil War (1861–65) he was chosen colonel of the second Mississippi regiment and played a prominent part in the first battle of Manassas (Bull Run). He retired from service after he failed to be voted to take command of his regiment in 1862, but soon "under a commission from President Davis he raised a partisan regiment of calvary and returned to the service."[39]

After the surrender of the Confederate Army, W.C. Falkner began to practice law again with his partner, Judge J.W. Thompson, and their firm "prospered." After he retired, leaving the law practice in the hands of his son, W.C. Falkner began the project of the Ripley Railroad. He became "the president" of the railroad company and "succeeded in putting it on a paying basis." Though engaged in this enterprise, he was still interested in local politics. At this time, the American South was "still in part under the domination of the 'carpet-bag' régime," and W.C. Falkner "earnestly desired to see his section restored to its rightful place in the national government." In the presidential election of 1876, he was one of the Tilden electors and "made an active canvass" for Tilden.[40]

W.C. Falkner was a charitable person, "the friend of the widow and the father of the fatherless in his community," as well as "a loving and faithful husband and father, and offered to all his children those educational advantages that had been denied him." He was also the author of several books, *The White Rose of Memphis*, *The Little Brick Church*, and *Rapid Ramblings in Europe*. W.C. Falkner was elected to the state legislature from Tippah County on November 5, 1889, and on the same day was shot on the streets of Ripley, which "closed suddenly and tragically" his career. His death was "mourned by multitudes who knew him throughout the State." Bondurant describes W.C. Falkner as a brilliant local figure who established his fame by following the typical rags-to-riches story. W.C. Falkner was a brave soldier in two wars and a successful figure in both his community and his literary works. This biography illustrates how the legendary image of W.C. Falkner was portrayed in the community after his death.[41]

W.C. Falkner's honorable military career was enforced in the tales of his friends and posterity. As Joseph Blotner writes, W.C. Falkner's first son, J.W.T. Falkner, periodically sponsored a reunion of W.C. Falkner's partisan rangers. Three brothers, including William Faulkner, gathered to hear memories of their brave ancestor during the Civil War. One veteran of W.C. Falkner's regiment recalled the danger and hardship of the regiment's campaign, but still boasted of the honor to be under W.C. Falkner: "with his own version of the rhetoric that had inspired the whole South in those long-gone days," the veteran "asserted that his service and sacrifice were a privilege of the highest order."[42] Bondurant's short biography is a summary of the local Southerners' memories of a typical hero.

The following biographical studies of W.C. Falkner, however, add information that dampens Bondurant's admiration. In his introduction for *The White Rose of Memphis*, Cantwell evaluated W.C. Falkner's *The White Rose of Memphis* as "a typical Southern romance" and included more precise biological information and its relationship to W.C. Falkner's fiction: W.C. Falkner killed two citizens—Robert Hindman and a man named Morris, "one of Hindman's partisans"—in 1849 and 1851. In both cases he was acquitted. The first was judged "a clear case of self-defense," and the second included a trial among "an intense pitch of excitement" in the town. As a result of these murder cases, W.C. Falkner was "condemned in the public mind as a murderer" and "when he walked down the street, he found deadly foes." Cantwell assumes these cases and the community's feelings are reflected in the detective's accusation of the protagonist in *The White Rose of Memphis*.[43] Cantwell writes a dramatic but dismal depiction of W.C. Falkner's tragic death:

> In 1889, while he [W.C. Falkner] was in New York, he was nominated to the Mississippi legislature, and was elected. On election night, November 5, 1889, after his victory had become assured, he walked to the public square at Ripley and was shot and killed while he was talking with a friend about sawing some timber. He had stopped directly below the office of J.H. Thurmond, a business rival and a very wealthy man, once his partner in the railroad, on the west side of the square. According to newspaper accounts of the time, Thurmond walked directly to Falkner, without saying anything, pulled his pistol from his pocket, and aimed it at the Colonel. Falkner said, "What do you mean, Dick? Don't shoot!" The bullet entered Colonel Falkner's mouth, breaking his jaw-bone, and lodged in the right side of his neck under the ear. Five men were standing near by, including Colonel Falkner's son-in-law. Sitting on the pavement, after his son-in-law had wiped the blood from his face, Colonel Falkner turned to Thurmond and asked, "Dick, what did you do it for?"[44]

Thurmond was accused in this case, but according to Cantwell, "Thurmond was acquitted in one of the most sensational trials in the history of Northern Mississippi."[45] This episode suggests that W.C. Falkner was not the completely honorable figure whose death was mourned by the whole community.

Duclos's study of W.C. Falkner's life and works follows Cantwell's description of the scene of Thurmond's attack, but give more details of what occurred afterward. After he was fatally shot, W.C. Falkner was carried to the home of his son-in-law rather than his own because his wife was in Memphis at the time, having left Ripley two weeks before with her daughter Effie to live in Memphis permanently. Rumor had it that W.C. Falkner and his wife "had not been getting along well for some time and that he had moved into his office to live, leaving the house to her." Duclos guesses that W.C. Falkner's "decision to remain in Ripley to become a political leader and his generally over-bearing personality may have alienated his wife's affection." Duclos's additional information suggests that W.C. Falkner's isolated condition was more serious inside his own family than in his town.[46]

Another description of the incident is given from a local source. In *The Falkner Feuds*, Hickerson vindicates Thurmond with "all the authentic records available" by claiming W.C. "*Falkner* was an arrogant (but intriguing) personality of many facets, while *Thurmond* was the quiet unassuming type of business man." Hickerson tells a different version of W.C. Falkner's death. On that day, W.C. Falkner had been drinking and making the rounds of the courthouse square with a number of his followers, and when he came to Thurmond's office he saw Thurmond sitting behind the open window and said something to him. Then W.C. "Falkner moved his hand toward his hip pocket in a manner indicating that he might be going after a gun; when he did, Thurmond, whose pistol was on the desk, fired, the bullet breaking his jawbone and ranging slightly upward." Hickerson's description seems to explain why Thurmond was acquitted in the trial. Or, Williamson's comment may more appropriately explain why the jury declared Thurmond not guilty: "[W.C.] Falkner alienated himself from his community by his independence, his willfulness, and his egoism. . . . Very few people in Tippah County would have killed Falkner as Thurmond did, but once it was done by him, many could approve the action."[47] W.C. Falkner's exceeding ambition appears to have caused two extreme reactions: some admired him as a heroic figure, while others loathed his arrogant personality. And the community's emotional reactions to W.C. Falkner continued after his death. While traveling in Italy, W.C. Falkner had ordered a marble statue of himself. The statue was brought to his hometown after his death, and his family erected it in the town cemetery in 1892. Apparently, some of the fingers on the statue's left hand were shot off by supporters of Thurmond.[48]

Separating legend from fact is very difficult in W.C. Falkner's biography. Arguing against Bondurant's column about W.C. Falkner, Cantwell, Duclos, and Hickerson claim other versions as the true stories. They question W.C. Falkner's reputation as a successful officer both in the US-Mexican War and in the Civil War, a generous slave owner and planter, an original founder of the Ripley Railroad, and a loving and faithful husband and father.

The details of W.C. Falkner's military careers are examined in the studies of Duclos, Hickerson, and Williamson. In his first stint in the military, W.C. Falkner rode away from his camp without permission from the commanding general in the field, and he was heavily wounded and retired during the US-Mexican War. In contrast, in his next military career in the Civil War, W.C. Falkner attained his most successful achievement in the field. In July 1861 at the Battle of Bull Run, he served effectively as colonel of the regiment.[49] These three biographical studies do not overlook the chance to admire W.C. Falkner's achievements during this battle. It seems unquestionable that Colonel W.C. Falkner was "one of the most distinguished volunteer officers" in the Civil War.[50] His next appointment as the personal commander of the

First Mississippi Partisan Rangers in 1862 was not official, but it was successful.[51] As Williamson argues, the rangers' object was to harass the enemy soldiers and use them as a source of supplies. The rangers were organized under a law that allowed them to be paid for munitions captured from the enemy. In short, "it was a universe constructed specifically for the talents and ambitions of William [C.] Falkner."[52] W.C. Falkner's military talent flourished during the Civil War, both in public and private service. After the Civil War ended, W.C. Falkner helped found Stonewall College, a postbellum project to revive the Ripley Female Academy that had been burned by federal troops in July 1864. For the formal opening of the college, he wrote and directed a play titled *The Lost Diamond*, a war-plus-romance melodrama.[53]

Reacting negatively to W.C. Falkner's image of a "humane" slave owner and planter,[54] Duclos and Williamson argue that W.C. Falkner was neither a member of the Southern great slaveholders nor of the great planters. Duclos says he was "frequently involved in small law suits over land and loans over the late forties and early fifties," and "was buying and selling whole and partial blocks of land in Ripley, and was the owner of several Negro slaves."[55] W.C. Falkner gained Negro slaves as part of his first wife's inheritance. On February 15, 1849, W.C. Falkner and his wife, Holland Pearce, sold for $1,200 four Negro slaves: Phillis, a thirty-two-year-old woman with three boys—John, age five; Joe, age three; and Pete, age one. On the same date, they purchased 320 acres of land for $2,200.[56] Williamson considers that "the Falkners were going into the farming business in a modest way"[57] but there is no record that he became a major player in plantation business. Rather, as Duclos suggests, during the post-US-Mexican-war period, though W.C. Falkner "was not *very* wealthy, he had begun to acquire fairly valuable property and was making numerous sound investments—buying parcels of land and later selling them at a profit."[58]

In spite of his idealization of pastoral life and portrayal of cruel slave trading in *The Little Brick Church*, W.C. Falkner raised their family's social status with land deals and slave trades. According to Williamson, the typical Southern "aristocracy" clans in Tippah County and the town of Ripley are the planters who have a dozen or more slaves.[59] However, W.C. Falkner did not have many slaves or a huge plantation in the antebellum era. W.C. Falkner was a war profiteer in the Civil War, and part of his cash seems to have come from his private robbery of the enemy. During the Reconstruction, W.C. Falkner became a member of the upper class in his local society. Williamson writes:

> During Reconstruction [W.C.] Falkner emerged as one of the several richest and most influential men in Tippah County. The richest of all, probably, was his associate Richard Thurmond. At least in the persons of these two men, the

> town came to dominate the country after the war, to turn the tables of hegemony from agriculture in the plantation slaveholding mode to commerce and finance. Neither man had invested directly in plantation slavery, and yet both were highly astute and aggressive businessmen who carefully built up and conserved impressive amounts of capital that they obviously did not invest in the Confederate war effort. When the plantation-slave system collapsed, each apparently made the most of a once in a lifetime opportunity to catapult himself out of the middling ranks to the forefront of wealth and power.[60]

W.C. Falkner's talent as a merchant prospered when the plantation system subsided after the Civil War. Although his imagination idealized the old plantation life, his aim upward in society affirms his talent with investing.

W.C. Falkner's main triumph of his businesses in the Reconstruction era was his ownership of the local Ripley Railroad. Bondurant perpetuated the myth by claiming the Ripley Railroad project "originated with Col. Falkner."[61] However, historical documents tell another story. Duclos and Hickerson both argue that W.C. Falkner did not take part in the original conception and early organization of the Ripley Railroad Company.[62] Quoting Andrew Brown's "The Railroad," Hickerson proves that "the Ripley Railroad Company was incorporated in 1856 . . . by 22 men, of whom [W.C.] Falkner was *not* one," though no work was done at that time.[63] It is in 1871 that W.C. Falkner appeared as a member of the thirty-six incorporators of the Ripley Railroad Company, which was newly authorized to build a railroad from Ripley to the nearest existing road. Sometime in 1871, W.C. Falkner was elected president of the new corporation.

One of the reasons for this project was a law in 1871 that said any standard railroad twenty-five miles or more in length built before September 1, 1872, would receive a bounty of $4,000 per mile. In 1871–72, a local stock company began work on the railroad. Despite the fact that the road was not standard gauge, was laid with thirty-six-pound rail instead of the fifty-six-pound steel required by law, and was only twenty-one miles long, the Board of Railway Commissioners accepted it as eligible for the subsidy payments. Of all these developments, however, there is no real evidence to prove W.C. Falkner's importance. The only recorded contribution of W.C. Falkner to the company was that in June 1872 he invested $500 to further the progress of the road. For this small investment, he received one first-class ticket that was good for fifty years. Duclos writes that similar tickets were issued to his wife and "to each one of their children then living or yet to be born, to each daughter-in-law or son-in-law or grandchild then living or yet to be born."[64] Although W.C. Falkner was president of the company at the time, it seems he did not contribute to the company until 1872.[65]

W.C. Falkner brought his dream of expansion to the Ripley Railroad Company. Before W.C. Falkner's ownership, the railroad bonds were bought in

1874 by R.J. Thurmond, C.L. Harris, and Chesley Hines. Thurmond had a two-thirds interest and the others a third. Later, Thurmond sold half of his bonds to W.C. Falkner. In 1886, W.C. Falkner bought Thurmond's remaining bonds. The railroad was sold at auction in 1889 and was purchased for $134,000 by W.C. Falkner, in partnership with C.L. Harris and Chesley Hines' heirs. Additionally, W.C. Falkner bought a railroad called "the Gulf and Ship Island," whose stretch of the road was "between Ripley and Pontotoc," in order to create "a bigger and newer corporation."[66] He was not the original founder of the corporation, but it is still true that the Ripley Railroad Company is an example of W.C. Falkner's ambitious rise in local society.[67]

W.C. Falkner also extended his family. Joel Williamson's study reveals that W.C. Falkner had children with a woman who was once his slave—his "shadow family."[68] W.C. Falkner's several descendants from 1848 to 1874 include his second family's ones. He gave his family name to all of his descendants and helped them go to school. But his eagerness for the extended prosperity damaged his white family's relationship, and this procreation seemed to increase his isolated condition in his white family. As Williamson argues, "there appeared to be a rising separation between William [C.] Falkner and his wife." In the middle of the 1880s, W.C. Falkner spent most of his time living in a hotel. He built an Italian villa in 1884 after returning from a trip to Europe, but he was "seemingly alone in his Italian Villa."[69] In the dedication of *Rapid Ramblings in Europe*, W.C. Falkner told readers that his legitimate white daughter, Roy, wrote a letter to her father asking him to "hurry home" because "Mamma is crying her eyes out" about him and "I am so very lonesome without you" (RRE 3). He ended his travel record with the band music of "Home, Sweet Home" (RRE 566). However, W.C. Falkner's image of a loving husband and father does not promise a happy relationship with his family members in later years.

W.C. Falkner, who was actually a man of commerce and finance in the New South, demanded to be remembered as a wealthy planter in the Old South. William Van O'Connor argues:

> Commentators on the Jefferson saga sometimes imply that Faulkner's antecedents were wealthy planters. This obviously is not true. Colonel William C. Falkner was one of the "new men," and J.W.T. Falkner, whatever his desires or private allegiances may have been, was involved with the political fortunes of the small tenant farmers, the very people who had in effect been disenfranchised during the old order.[70]

This controversial connection between reality and imagination suggests that, in the process of the American South's economic modernization, a Southerner's imperial desire is retained by imagining the ideal man in the old plantation culture—the glory of military operation, the loving father and husband,

and the generous slave master. In the first half of the nineteenth century, the Southern plantation system remains "profitable because of the demand for cotton," and the white planters' imperial will is locally exercised by their control of slaves and "a rigid line of color and status between the races."[71] However, the local planters' struggle to keep this regime results in the defeat of the Confederate army in the Civil War and the region's economic decline. The Southerners' local model of the plantation culture and its dominance only remains in their imaginations. W.C. Falkner's biographical studies prove that his legendary image emerged from his talent in merging his community's romantic imagination with his life story. His real life record, however, clarifies that the images of the large plantation's owner, the original founder of a railroad, the faithful husband, and the loving father of a decent white family are far from fact. It seems that his romantic literary works do a lot to enhance W.C. Falkner's historical image. Comparing W.C. Falkner's works with those of his great-grandson is useful not only to see how W.C. Falkner's romantic and fictitious imagination is ironically used in Faulkner's narratives, but also to see how Faulkner insinuates and criticizes the effect of W.C. Falkner's imagination to the Southern people's modern lives.

THEMATIC COMPARISON BETWEEN *THE SIEGE OF MONTEREY* AND *ABSALOM, ABSALOM!*

This part compares W.C. Falkner's first published poetry book, *The Siege of Monterey*, with Faulkner's *Absalom, Absalom!* to explore how Faulkner reinterprets his great-grandfather's romantic imagination. Faulkner arguably connects Sutpen's appearance in Yoknapatawpha with W.C. Falkner's ambitious publication of his first literary work. W.C. Falkner (b. 1825) published his first poem, *The Siege of Monterey*, when he was around twenty-six years old (1851). Thomas Sutpen (b. 1807) appeared in Yoknapatawpha County when he was around twenty-six years old (1833). When Faulkner (b. 1897) was twenty-six years old in 1923, he "wrote a book" and discovered that his "doom, fate, was to keep on writing books." Faulkner's short biographical profile to the publisher of *The Marble Faun* on September 9, 1924 reveals he was still stuck in W.C. Falkner's shadow. Faulkner writes, "Born in Mississippi in 1897. Great-grandson of Col. W.C. Faulkner [*sic*], C.S.A., author of 'The White Rose of Memphis,' 'Rapid Ramblings in Europe,' etc."[72] This episode suggests his obsessive connection with his great-grandfather's literary career. Faulkner thought about W.C. Falkner's first publishing age when he published his first poem, *The Marble Faun*, and W.C. Falkner's romantic appearance in the Southern imagined community when he composed Sutpen's life.

It might be true that Sutpen's model is W.C. Falkner. As Duclos argues, "Faulkner attributes to Sutpen certain characteristics which seem to be inspired by the legend of Colonel [W.C.] Falkner. Sutpen came into Jefferson penniless and raised himself in society by perseverance, arrogance, and struggle. The boyhood dream of Colonel [W.C.] Falkner is echoed in the youthful 'design' of Sutpen."[73] Duclos's argument suggests young Sutpen is an echo of young W.C. Falkner's ambitious prospect for success. However, it might be the more important fact that W.C. Falkner published his first poem, *The Siege of Monterey*, almost at the same age as Sutpen is when he appears in Yoknapatawpha County. Comparing W.C. Falkner's poetic talent that predicts the author's future prosperity with Sutpen's ability to create his own ambitious design, this study explores how W.C. Falkner's romantic monologism is criticized in Faulkner's dialogic space.

The first section of the following argument examines how W.C. Faulkner uses monologism to create his literary world in *The Siege of Monterey*. The second section compares *The Siege of Monterey* with *Absalom, Absalom!* to explore how Thomas Sutpen's life is a reflection of W.C. Falkner's ambitious prospect for the future, and how young two narrators in *Absalom, Absalom!* react to this gravitating personality. In Faulkner's work, Thomas Sutpen is depicted as a typical embodiment of the monologic imagination of W.C. Falkner's literary world.

A Hero's Monologic Dominance of Life, Love, and Race: *The Siege of Monterey*

W.C. Falkner's first poetic work reveals the author's monologic identification with his romantic heroes' brave lives. *The Siege of Monterey* (1851) is an epic of the US-Mexican War (1846–48), which contains an autobiographical preface in order to synthesize the author's voice with the epic's heroes. This monologism gives a total perspective to this badly composed poem, which has eight cantos. After his biographical preface, W.C. Falkner writes his own biography in the first canto, and then expounds upon the American army's glory in the second canto. From the third canto to the eighth one, he praises the romance of the hero and heroine of the Mexican army, Bibo and Isabel. The protagonists of these three parts are different, but all these cantos are the author's valiant, self-justifying narrative to excuse his Ripley murder in 1849 and to foretell his fortunate future. Every hero of this epic poem is depicted as gallant, self-defensive, and honorable. *The Siege of Monterey* is written to defend the author's murder case as a heroic deed that will lead him to a prosperous life.

The Siege of Monterey is a romantic author's literary apology and promise for the future to overcome his sinful mistake of the past. W.C. Falkner uses its preface to explain his cornered condition in his community, and then writes the first canto to defend against the blame of "my foes" (SM 3). His aggressive tone continues to the second canto, in which a historical figure, Major General Zachary Taylor (1784–1850) of the US army, declares the commencement of the Battle of Monterrey on September 21, 1846, during the US-Mexican War. Then, from the third canto to the end, Bibo and Isabel appear as brave Mexican soldiers in the siege of Monterey. Bibo and Isabel finally marry and make a wealthy life in England. The main characters are varied in this poem: in the first part, it is obvious that W.C. Falkner is the protagonist; Major General Taylor commands the second part; and Bibo and Isabel are the protagonists for the rest of the poem. And yet these characters share the same militant mood against their foes to strongly suggest these characters are the alter egos of the author.

The author's identifications with these main characters are achieved through their aggressive attitudes. In his preface, W.C. Falkner challenges his enemies: "I say to any of my foes, if I have wronged you, come up like men and demand satisfaction" (SM vi). In the first canto, he warns to his "foes" that "you shall to your sorrow know, / 'Who conquers me shall find a stubborn foe.'" (SM 3–4). Major General Taylor says, "the gallant warrior firmly faces his foes" in the second canto (SM 17). And Bibo is described as a hero who dreads "no deadly foe" (SM 42). Though Bibo is a colonial Mexican who fights against the American army, W.C. Falkner still identifies himself with this character, suggesting that both of them share the same aggression as a brave man. All of these protagonists show a common hostility toward their foes, even if their enemies are different.

This poem's protagonists, the young Mississippian author, the US military man, and the Mexican soldier, are identified through their common violent passion. By integrating these characters, W.C. Falkner creates his romanticized poetic world that can conquer its enemy. As Bakhtin argues, in romantic poetry, the poet strips "all aspects of language of the intentions and accents of other people" and destroys "all traces of social heteroglossia and diversity of language" to achieve "a tension-filled unity of language" (DN 298). In this process to create a monologic world, "the unitary poetic language" is completed (DN 298). In contrast, in the modern novel, the author "welcomes the heteroglossia and language diversity of the literary and extraliterary language into his own work not only not weakening them but even intensifying them" (DN 298). In his cantos, W.C. Falkner disregards the diversity of the world and unifies the plural protagonists' worldviews with a heroic tone.

W.C. Falkner's monologic poetry makes a strong contrast with the dialogic structure of his great-grandson's first poem. Faulkner, who begins his poetic career with a poem titled *The Marble Faun* (1924), completes his first world that contains plural perspectives. This difference might be one of the reasons

Faulkner fails to be a successful poet. In his famous comment on his talent, he says, "I'm a failed poet. Maybe every novelist wants to write poetry first, finds he can't and then tries the short story which is the most demanding form after poetry. And failing at that, only then does he take up novel writing" (LG 238). As his poetry shows, Faulkner cannot stick to one voice—he always wants to have plural voices and worldviews. In *The Marble Faun*, the narrator, who is a marble statue, is isolated from the world and observes the rich and free garden around him, where the other character, Pan, resides. Faulkner depicts two worlds in one poem, and there is no element to unify these two worlds in one perspective. His other poetry work, *A Green Bough* (1933), is a collection of poems. In this work, every poem has a different narrator, and each seems to be a poetic version of a short story. In these two poetry works, Faulkner reveals his modern preference for dialogic structure. Contrasting W.C. Falkner's talent and Faulkner's, readers can see that W.C. Falkner is essentially a romantic poet who wrote prose, and his great-grandson is essentially a novelist who failed to be a romantic poet.

W.C. Falkner reveals his romantic, monologic talent in his first poem. The characters share the same sense of belonging through love, violence, and death. This sense of strong unity goes so far as to allow incestuous, homosexual innuendo. In an episode on the battlefield, two brothers are killed at once and "both closed their eyes in the shades of death" (SM 91). To depict the dead brothers' close bond, W.C. Falkner boldly compares their relationship to marriage:

Both brothers lay closed in each other's arms,
  Like passionate lovers their bodies lay
As if feasting on each other's charms,
  Like lovers do soon after their marriage day.
Their eyes seem closed in slumber sweet,
  Their blood mingles together on the sod,
Their souls in each other's embrace retreat
  To the shining mercy-seat of God. (SM 91)

W.C. Falkner uses incestuous, homosexual implications to insist on the characters' unity. In his strong desire for the unity, his imagination ignores cultural taboos.

From this poem's preface to the second canto, the young author's excessive demand for the romantic unification with his imagination and his characters pushes on his writing, engulfing himself and Major General Taylor in the same passionate turbulence of the desire of unification and antipathy against foes without any definite goal. However, when W.C. Falkner narrates Bibo and Isabel's romantic epic from the third canto to the last, he finally finds this poem's main theme: the violent noble race can protect its belongings from foes and achieve a wealthy life. The hero and heroine's imperial impulse to fight against their enemies is the most important element to prove their race's nobleness. Isabel, who disguises herself as a young soldier "to protect" Bibo (SM 42), demands to go along with Bibo in the field. The narrator says her beautiful features are "an equal match for the heavenly race" (SM 38). On the other hand, Bibo is written as "a brave and noble boy" because "many a deadly foe did he destroy" (SM 105). These heroic characters' violent actions are used to prove their superiority or dominance over others. Bibo "dashes amid the thickest of his foes" (SM 77) to prove that he comes "of a very noble race" (SM 105), and the way he goes is "paved with gore" and full of "ponderous heaps of dying foes" (SM 77). And Isabel "lopp[s] off" her enemy's head, who falls "dying near her feet" (SM 98). Bibo discovers his lover's beauty through their violent actions in the field. When Isabel tells Bibo," Side by side let us combat our foes" (SM 95), Bibo finds "angelic beauty" fills "her eyes" (SM 96). This connection between the heroine's violent urge and her lovability is a typical pattern of W.C. Falkner's imagination. Isabel's father, Don Pedro, does "love his daughter . . . / For many foes did she slaughter" (SM 105). In their heroic deeds, Bibo and Isabel's violence is proof of their nobility and lovability.

W.C. Falkner learns his romantic narrative's tone from Sir Walter Scott's literature. As V.W. Brooks argues, in the antebellum era many Southerners shared the same literary taste—"what Mark Twain called the 'Sir Walter disease.'"[74] Scott's chivalric romanticism pervaded Southern literature in the nineteenth and early twentieth centuries. The Southern lifestyle is mingled with Scott's fictional romance. This characteristic sense of assimilation between fiction and reality creates the romantic South's narcissistic self-sufficiency, which does not feel the need to explain itself to foreign audiences. In his essay, "Golden Apples" (1937), Stark Young argues that traditional Southern literature "is apt to be too complete within itself and too contented to feel any need or urge to preach to others outside."[75] Young considers that Southerners learn this sense of self-sufficient romanticism from Scott's literature:

> In literature the tradition of the romantic South of the plantation came to reign. . . . How far people believed in the tradition in all its details, how far the pattern was accepted, I do not know, what with life following art as well as art following life. But people did by thousands read books and sing songs composed out of that cavalcade of romantic material, and tears were shed and royalties earned on this nostalgic, and even Walter Scott, basis.[76]

The romantic Southerners find their ideal models in Scott's works that will lead readers to anachronic monologism. As Joep Leerssen writes, the plots of Scott's novels are romantic adventures that involve "a new interest in the past, preferably the medieval past" of Western Europe.[77] Scott's *Ivanhoe* (1820) is a typical novel of romantic medievalism, in which the cavalier hero survives in the battlefield and keeps love in his heart for "the Saxon Princess," Lady Rowena, in twelfth-century England; in heroic tone, the novel's narrator is almost identified with Scott's voice and depicts "a field slippery with blood, and encumbered with broken armour and the bodies of slain and wounded horses" and the honor proclaimed "to the brave" and glory "to the victor."[78] W.C. Falkner's little story of "*love*, *war*, *blood*, lust, and glory" (SM 3) uses Scott's technique to synthesize the hero's violent actions and utterances with the author's ideal self in the narrative's highly aggressive tone.

Under Scott's influence, W.C. Falkner's monologism becomes a narrative that links his arrogant worldview with the main characters' lives. As Bakhtin argues, the romantic author seeks to identify his own voice with characters to certify his own worldview, and his own words are placed in the mouth of any other of his characters "with equal success" (PDP 79). As a result, the monologic narration seemingly bridges the distance between the author and the characters. In the first and second cantos, the narrator is identified as W.C. Falkner by its heroism, and he is also identified with Bibo and Isabel by the romantic atmosphere. Using Scott's broadly accepted plot, W.C. Falkner's monologism allows the author to expect a revival of his own honor when his romantic hero attains his prosperity at the end of the poem.

In W.C. Falkner's poem, the author's autobiographical aspect is reflected in the hero's aggressive voice. As Duclos correctly argues, "As poetry, it [*The Siege of Monterey*] is generally bad; as the narration of an historic event, it is far from complete; but as an insight into [W.C.] Falkner's life and thought, it is in many ways the most significant of his many literary works."[79] In a way, this poem contains W.C. Falkner's private confession. To connect himself with the hero in the epic, W.C. Falkner reveals why he wants to create this poem in the preface: he confesses that he has serious trouble with a "Mr. H."[80] Blotner informs "Mr. H." is Robert Hindman, who was a private in W.C. Falkner's company. In the spring of 1949, serious trouble with an unfortunate result happens between W.C. Falkner and Hindman. Blotner writes:

> The Hindmans were a brave, vigorous, and able family also known to be high-strung, reckless, and aggressive. Now Robert wanted badly to belong to another organization in which [W.C.] Falkner had some influence, the Knights of Temperance. [W.C.] Falkner spoke on behalf of Hindman's admission, but someone told Hindman just the opposite, and he swore he would kill [W.C.] Falkner. The young lawyer went out to the Hindman place, and when he denied the blackballing, Hindman angrily replied, "You're a damned liar," and pulled out a revolver. While they struggled it misfired twice. As it clicked a third time,

> [W.C.] Falkner drove home a knife and Hindman fell dead. It was clear that [W.C.] Falkner had acted in self-defense and when the case eventually came to trial he was acquitted.[81]

After this trouble with Hindman, W.C. Falkner "turned back to his law practice and the growing [W.C.] Falkner Farms, but he was harassed by the Hindmans and felt himself in constant danger."[82] W.C. Falkner's stressful struggle with Hindman is reflected in *The Siege of Monterey*.

In the preface of *The Siege of Monterey*, W.C. Falkner outlines the Hindman's harassment. While traveling, shortly after killing Hindman, he came to the house of an old lady and inquired if he could remain at her house during the night. However, when she heard his name, she furiously rejected him because she knew he had murdered Hindman. To prove his murder was in self-defense, he read the testimony of the case. W.C. Falkner found his reading gave a dramatic change to the old lady's attitude:

> The old lady listened very attentively, until I finished reading, then sprang to her feet and begged to be forgiven for the treatment which I had received. I told her she had done nothing wrong, that her author was the criminal. In less than twenty minutes, she prepared an excellent supper, and treated me with as much kindness as if I had been her own son. (SM viii)

After this episode, W.C. Falkner concludes, "I hope I shall make no new enemies by my rhymes, for God knows I have a good supply on hand at present" (SM ix). This preface suggests what the author intends to do in his poem: justify himself to make the community's opinion preferable to him and to make people believe his promising talent. In the end of this poem, Bibo and Isabel obtain a happy life with great fortune in England. Hoping his poem has not made readers his "foes" (SM 131), W.C. Falkner's imagination—which is propelled by his desire to unify the world with his imperial will—goes easily beyond the nation's borders and overlooks any cultural and political differences to suggest that the author, as demonstrated by his hero, will overcome his struggle with his enemies and gain a successful future.

The romance of the hero and heroine in the latter six cantos of this poem is identified with the author's real romance. In the first canto, W.C. Falkner suggests that the author himself is "in love clear up to the nose" (SM 7). Blotner considers that W.C. Falkner is confessing his love for his second wife, Lizzie Vance: "the poet celebrated his love for Lizzie Vance and also the benefits of the married state." The hero's wounds show the identification between the author and the protagonist as well. According to Blotner, during the US-Mexican War, W.C. Falkner's left foot was "shattered by one ball and the first joints of three fingers of his left hand blown away by another."[83] In response to this experience, in *The Siege of the Monterey*, Bibo gains also a serious wound in the process of escaping from the enemy.

W.C. Falkner's wounds in the US-Mexican War are far from the honorable. Hickerson argues that some people think the Tippah Guards, which W.C. Falkner was part of, did not achieve any honorable accomplishment in the war: "They had no battle experience, but suffered heavily from smallpox and diarrhea. . . . He [W.C. Falkner] had been shot in the left foot and the fingers of his left hand. His enemies in Ripley declared that he suffered these wounds in a brawl."[84] W.C. Falkner's wounds seem to be the results of shameful deeds. This fact is reflected in Bibo's wound. In the end of the poem, in the process of escaping from Mexico to England, Bibo gained the serious wound on the ship when he fell from the high towering mast against the deck, "very near dislocating his neck" and "breaking a pair of ribs and his left arm" (SM 125). Bibo's wound was an accident—not an honorable mark from the battlefield. At the end of the story, after he gained an inheritance from his father, Bibo's sitting chair is "a fine self-propeller" (SM 131). The protagonist's wounded condition and his wheelchair parallel the author's accidentally lost fingertips and wounded left foot.

This identification between the author's life and the hero's one strongly effects the racial category of the hero and heroine, who both share the same racial features as W.C. Falkner and his fiancée. Bebo and Isabel of *The Siege of Monterey* are Mexican, but their shared racial features insist on a connection between the author's couple and Bibo's one. Bibo's features not only tell of his superiority, but seem to suggest the author's narcissistic self-image:

> He [Bibo] was beautiful, generous, and kind,
> He possessed a tender heart and witty mind,
> . . . .
> A good head, full of generous knowledge,
> Well cultivated at a Spanish college;
> Features fair and smooth, figure neat,
> Roman nose, blue eyes, and little feet—
> The only son of Don Zebboneezer—
> A heart as brave as Julius Cæsar;
> Skin as white as polished alabaster—
> Just such as young ladies sigh after. (SM 37)

The author swears "by the most holy truth" that Bibo is "a most lovely youth" (SM 37). The author's celebration of the hero's beauty enforces this poem's racial preference. Isabel is also an ideal beauty:

Isabel was Don Pedro's only daughter,
. . . .
The damsel's skin was as white as snow,
Her eyes like beauteous pebbles show,
Which shine beneath the sparkling brook;
'T was heaven itself on her check to look—
Part of heaven shone in her lovely eye,
The balance a short space below them lie;
Her beauteous charms and lovely face
Were an equal match for the heavenly race. (SM 37–38)

Bibo and Isabel are European colonists' descendants who have white skin. By including the racial features of this couple, the author creates an ethnological homogeneity between Bibo and Isabel and strengthens their identification with the author and his fiancée.

This identification also extends to their descendant, though this expectation comes in a threatening way. Their racial homogeneity is mentioned again to promise their descendant's continuous prosperity, but the two stanzas near the end of this poem depict the sudden surge of doubt regarding the hero's descendants' racial category. After their marriage, the hero's happiness peaks with the birth of his child, but Bibo feels the sudden need to make sure of his wife's fidelity by checking their child's racial features—in the past, a man, Don Gusto Conner, said "one single slanderous repute" on Isabel's fidelity (SM 117). His wife's chastity must be proven by their child's whiteness:

Bibo's cup of joy was now full to the very brim,
  Blessed with a beauteous and lovely wife,
Whose thought was of him, and only him,
  For she loved him as she did her life,
And humored him in his every whim,
  And never engaged in disgraceful strife;
And to crown his unexpected joy,
Isabel gave birth to a beauteous boy.
. . . .
Which set Bibo's brain in a perfect whirl,
  With pure love and paternal delight;
Though he wished it had been a girl,
  Yet his eyes with joy were exceedingly bright.
He asked the midwife "if his hair did curl,
  And was his complexion very white;"
To which inquiry, she answered "yes;"
"There's no doubt but he's mine then, I guess.["] (SM 129)

In this last stage, the racial appearance of the hero and heroine's child is a serious factor to make certain the legitimacy of his inheritance. Bibo, who has not forgotten the rumor of his wife's infidelity, asks the midwife about the racial feature of Isabel's child. Even in the peak of his love, the hero could not forget the possibility of his wife having a racially different lover. To accept his son as legitimate, the whiteness and resemblance is considered a good excuse.

The hero's arbitrary definition of his son's race does not need any conversation with his wife. This matter suggests what kind of monologic voice dominates W.C. Falkner's imagined world. The most important point here is the author's identification with the hero. Because the hero represents the author's sympathetic will, Bibo's monologic discourse can define the legitimacy of his son by himself. Bibo's identity presupposes his wife's complete obedience to his will, and he feels no need to tell her of his suspicion about their son's racial category. The fact does not come from the dialogue, but from the hero's monologic voice which represents the author's will. Race is not terribly important in W.C. Falkner's work because it is not decided by objective facts but by the white male's monologic decision and utterance.

What Bibo exercises here is not his objective examination of a child's race but his monologic dominance over his child's racial category in his narrow community. In W.C. Falkner's imagination, one of the most effective powers of monologism is the materialization of the hero's voice. According to Bakhtin, in monologic speech, an individual utters definitive words according to whole objects. The hero's—and, consequently, the author's—words define who is the enemy, who is the friend, and what happens now. Monologism can create reality by its narrative: "With a monologic approach (in its extreme or pure form) *another person* remains wholly and merely an *object* of consciousness. . . . Monologue manages without the other, and therefore to some degree materializes all reality. Monologue pretends to be the *ultimate word*. It closes down the represented world and represented persons" (PDP 292–93). W.C. Falkner's poem uses the descendant's racial problem as a dramatic factor to reaffirm the hero's identification with the author. The effect of monologic power is demonstrated by the decision of child's race. When the hero says his son is "mine," he actually creates his son's racial purity (SM 129). The author unifies his voice with the hero's and allows his hero's monologue to determine his son's legitimacy without any objective evidence.

At the end of this epic poem, the hero's father, Don Zebonezer, dies and leaves his large fortune to his son. Bibo enjoys his luxurious life as the conclusion of this epic:

Bibo now built a very spacious hall,
  For he now had plenty of persuasive gold;
Silver burnished was his polished wall;
  Servants he had with numbers untold;
Notwithstanding he had drank his share of gall,
  His cup of bliss was as full now as it could hold;
His sitting chair was a fine self-propeller;
Choice wines filled his spacious cellar. (SM 131)

Bibo's happy life concludes with a great fortune, his white wife, and his white son. This satisfaction is celebrated by the hero's (and the author's) self-content words. Yet the hero's wounded body, which is supported by his sitting chair that is "a fine self-propeller," suggests some fatally different worldviews might be possible on this protagonists' glorious ending.

In *The Siege of Monterey*, W.C. Falkner aims to attain two purposes at once. The first purpose is the author's self-justification of his battle against his enemies in real life. His foes are the Hindman family and the "author" of his bad reputation (SM viii). His poem intends to invalidate their persecution of W.C. Falkner for his murder. The second purpose is to narrate the romantic hero's successful marriage with the heroine as a predictive model for the author's remarriage with his childhood sweetheart, Elizabeth Vance, and his future prosperity and progeny. These two purposes are accomplished by the author's identification with the poem's hero, Bibo, who kills many foes in the battlefield and gains a beautiful wife, a white son, and a large fortune in his monologic world.

Challenges to Monologism in *Absalom, Absalom!*

In *The Siege of Monterey*, W.C. Falkner's high ambition for the future is reflected in the romantic hero's prosperity and formation of the white decent family in England. The hero (a portrayal of the author) becomes his family's authoritative voice that decides his child's racial category. This monologic feature of W.C. Falkner's narrative is used to ultimately celebrate the hero's (and the author's) absolute dominance of his narrow community. In contrast to *The Siege of Monterey*, in Faulkner's *Absalom, Absalom!* plural narrators, Rosa Coldfield, Quentin's father, Quentin and Shreve, dialogically describes Thomas Sutpen's failure to form his ideal self-centered world in Yoknapatawpha County. Sutpen's life is totally dominated by his own ambitious design. Sutpen was born in Virginia area as a son of a poor white in 1807, and he began to conceive his life-plan to achieve the typical successful life as a plantation owner in the American South when he was around fourteen years old. Young Sutpen went to Haiti to establish his material wealth, but he failed in his marriage with a sugar planter's daughter. He abandoned his first wife Eulalia Bon and his son Charles Bon, and left Haiti with some slaves. He returned to America and appeared in Yoknapatawpha County, Mississippi, in 1833 to acquire a hundred square miles of bottomland. His efforts turned his land into a huge plantation, which was called "Sutpen's Hundred" (AA 4). He remarried with a good white family's daughter and gained his legitimate son, Henry, and his daughter, Judith. After the defeat of the Confederate army, Sutpen's reduced circumstances and tragic death meant his design failed. Nevertheless, Sutpen's ambition still has a strong influence because his iron will to create his own future has not lost its attraction for modern people. What Faulkner dramatizes in *Absalom, Absalom!* is the modern narrators' struggle to form their dialogical space by resisting Sutpen's monologism, which strongly influences people's lives in the twentieth century.

In the second half of *Absalom, Absalom!*, two young Harvard University students, Quentin and Shreve, try to make sense of the reasons for the tragic fates of Sutpen and his descendants. In the process of their dialogical narratives, these two young narrators reveal their worldviews and how their preferences influence their narratives. At the end of this novel, Quentin reveals he is attracted to Sutpen's monologic worldview, while Shreve is critical of the Southerners' self-centered attitudes. Quentin and Shreve explore the tragic fates of Sutpen and his descendants to examine the adequacy of their own worldviews.

At the beginning of their dialogue, Quentin mainly talks about Sutpen's life following information he heard from Rosa and his father. He narrates how Sutpen's first illegitimate son, Charles Bon, appears in front of Henry as a university student without telling Henry they are half-brothers. Henry is strongly attracted to Bon and introduces Bon to his sister, Judice. Judice becomes Bon's sweetheart, a mirror image of Henry's homosexual desire for Bon. Quentin narrates this confused, homosexual, incestuous relationship to make his narration reflect his own incestuous desire for his sister, Caddy. Quentin feels Sutpen prevented Bon and Judice's marriage because of its incestuous nature. However, as the dialogue proceeds, Shreve overtakes Quentin's narrative and gives his own interpretation of the story. Shreve suspects that Sutpen's motivation to reject his first son, Bon, was caused by the fact that Bon racially inherited the African American blood. The difference between these two young narrators' interpretations of historical fact reveals that their narratives reflect their own personal worldviews.

Quentin's attitude toward Sutpen's life makes a strong contrast to Shreve's. Quentin is the white Southern family's elite who believes that one worldview will be inherited from the father to the son, and that plural narratives are to be tied to one authoritative narrative. While talking about Sutpen's life, Quentin finds that both he and Shreve are influenced by the interpretation of Quentin's father, and expects that their voices can be one to create their unified story. Quentin thinks

> *Maybe we are both Father. Maybe nothing ever happens once and is finished. Maybe happen is never once but like ripples maybe on water after the pebble sinks, the ripples moving on, spreading, the pool attached by a narrow umbilical water-cord to the next pool which the first pool feeds, has fed, did feed, let this second pool contain a different temperature of water, a different molecularity of having seen, felt, remembered, reflect in a different tone the infinite unchanging sky, it doesn't matter: that pebble's watery echo whose fall it did not even see moves across its surface too at the original ripple-space, to the old ineradicable rhythm* thinking *Yes, we are both Father. Or maybe Father and I are both Shreve, maybe it took Father and me both to make Shreve or Shreve and me both to make Father or maybe Thomas Sutpen to make all of us.* (AA 210)

In his imagination, it is revealed that Quentin is caught in the gravity of the monologic narrative of the American South. He hopes their whole, unstable narratives will eventually reflect the truth of Thomas Sutpen's life, as if the "*ripples*" caused by the pebble will stabilize under "*the infinite unchanging sky.*" Quentin expects to share the same perspective on the tale of Sutpen's family with Shreve, as if they were one "*Father.*"

Quentin expects that his and Shreve's narratives will be unified to totally understand the story of Sutpen's life and his family; however, Shreve becomes so excited by what he has heard that he creates another version. Shreve's attitude reveals that their dialogic relationship creates differences in people's voices and worldviews, and his excitement for the narrative continues because of his re-accentuation of the authoritative discourse. Even if their narratives are once identified under the influence of Quentin's father, this agreement cannot continue eternally. A narrative's unified vision is stratified by centrifugal force that is caused by another narrator's individual perspective. Continuation of dialogue between individuals inevitably creates fundamental differentiation in their interpretations of the world.

It is true that Quentin and Shreve establish an interpersonal voice that has not existed before, but it is not the birth of a new authoritative discourse—it is the discovery of the relativity of plural narratives. As the author, Faulkner comments that the birth of Quentin and Shreve's interpersonal narrative disproves the reliability of history: "both thinking as one, the voice which happened to be speaking the thought only the thinking become audible, vocal; the two of them creating between them, out of the rag-tag and bob-ends of old tales and talking, people who perhaps had never existed at all anywhere, who, shadows, were shadows not of flesh and blood which had lived and died but shadows in turn of what were (to one of them at least, to Shreve) shades too" (AA 243). Faulkner suggests that, when Shreve and Quentin recreated the narrative with their intersubjective voice, their narrative fabricated the "people who perhaps had never existed at all anywhere," as if they cooperatively projected other illusions on the screen. Faulkner uses Shreve and Quentin's intersubjective voice to relativize the past as another fiction that has been created by someone else. Shreve, an international student from Canada, can accept this attitude and enjoy the American South's past as fiction. His tale of Sutpen's life is an examination of the symbolic order that supports American South's social system, not the fact.

In the latter part of this novel, Shreve exercises his own right to be a father of his tale. As he finds that the racial identity is an important factor of the American South's social structure, he guesses the racial belonging of Bon as a secret matter and depicts Bon as part African American. With this additional element of racial belonging, Shreve creates his own tale by himself. Shreve guesses that Sutpen discovers his first wife is "*part negro*" upon meeting his firstborn son (AA 283), assuming the main motivation to obliterate his first son from his family is Sutpen's fear of "*miscegenation*" (AA 285). He dramatically depicts how Henry killed Bon to prevent "*the nigger*" (AA 286) from becoming a member of the family. As Marta Puxan argues in "Narrative Strategies on the Color Line: The Unreliable Narrator Shreve and Racial Ambiguity in Faulkner's *Absalom, Absalom!*" (2007), Shreve is "an unreliable narrator" who arbitrarily creates the story by himself, telling "Sutpen's story by establishing an order of events according to his personal interest, selecting some scenes and characters that had not been of such importance in previous versions of the story."[85] Puxan argues Shreve invented Bon's background to give a logical explanation for the gaps in this novel. As Noel Polk suggests, there is no evidence that Bon is actually part "black,"[86] but Shreve invents the idea of the fear of "miscegenation" to explain the reason for Sutpen's desertion of his first son.[87] Shreve reinterprets Quentin's obsessive, incestuous story into a race-obsessed tragedy. Puxan explains that one of the main parts of *Absaom, Absalom!*, especially the explanation of the motivation behind Henry's murder of Bon, is the product of Shreve's imagination.

In his examination of Sutpen's life, Shreve assumes the most fearful taboo for the white male Southerner is losing the power of his monologic dominance of his narrow community and the appearance of another formation of its order. This threatening possibility of the reformation of the small community appears as the fear of racial confusion within the family. At the end of his story, Shreve supposes Henry killed Bon because Sutpen told Henry that Bon's mother was "*part negro*" (AA 283). Shreve finds that in the traditional Southern society's social structure, the descendant's racial belonging was decided by the patriarch's monologic voice. Once Sutpen said Bon's mother was "*part negro*," his descendants considered she was "*part negro*." Shreve uses this social structure to explain the dramatic convert of Henry's attitude to

Bon. If Bon wanted to enter Sutpen's house as Henry's white brother, Henry would welcome Bon as his homogeneous double; however, after he heard Sutpen's monologic confession on Bon's racial belonging, Henry believed that Bon was part African American and shot Bon when Bon wanted to enter Sutpen's house as "*the nigger that's going to sleep with*" Henry's sister (AA 286). Shreve is excited about his interpretation, and asks Quentin whether his interpretation sounds true or not. Quentin says "yes" (AA 287), and finds his demand to create a unified worldview ironically encourages Shreve to tell another story of Sutpen's family. Finally Quentin understands that the total unification of the tale is only possible in an isolated narrator's monologue, and that there is no way to unify people under a history when people have dialogical relationship with the historical facts. The only way to stabilize one's own tale is to exclude all the other voices from the narrative. Quentin and Shreve can be equal partners to produce variations in their narratives, but this dialogic relationship's result cannot satisfy Quentin's demand for the unified world. Because of this discovery Quentin's suicidal attempt to be isolated from the world is established in this novel's end.

In contrast to Quentin's desire for the total unification of voices that support the decent family's homogeneity and Henry's incestuous demand, Shreve's mocking of white supremacy in the American South ends with a sarcastic conclusion that suggests the multiracial future of the American South. After understanding that people's racial belonging is controlled not by the father's faithful relation to his white wife but by his monologic narrative, Shreve expects that Jim Bond, one of Sutpen's descendants born between Charles Bon and his octoroon mistress, can be a symbol of miscegenation that will merge the other races to create uncertainty in racial distinction, even if one's appearance is white. Shreve predicts:

> I think that in time the Jim Bonds are going to conquer the western hemisphere. Of course it wont quite be in our time and of course as they spread toward the poles they will bleach out again like the rabbits and the birds do, so they wont show up so sharp against the snow. But it will still be Jim Bond; and so in a few thousand years, I who regard you will also have sprung from the loins of African kings. (AA 302)

Puxan interprets this part as "Shreve's terrible prediction about white supremacy and white purity" (552). It is true that Shreve believes in white supremacy, but it is also true that Shreve knows his narration contains an exaggerated irony regarding the uncertainty of racial distinction. In his narrative, the racial category is arbitrarily decided by Shreve's intentions and expectations for the future. In Shreve's narrative, Sutpen becomes a person who establishes his design expecting that his white descendants will inherit his dynasty and reproduce that design to be richer and richer. This design includes the presupposition that only white descendants can achieve future prosperity. Sutpen follows this presupposition and chooses Henry, not Bon, as his legitimate child. Sutpen's racial segregation of his descendants comes from his expectations for the future. Shreve criticizes this idiosyncratic presupposition re-accentuating the symbolic meaning of his white body, assuming that if he and Sutpen presuppose that the miscegenation will thrive over "to conquer the western hemisphere," it is highly possible that Sutpen accepted Bon and Jim Bond as his legitimate descendants and also Shreve's own racial distinction comes from this future expectation; therefore, he who regards Quentin "will also have sprung from the loins of African kings" (AA 302). Shreve's mocking narrative insists that current race distinctions are decided by the people's prospect for the future.

Naturalization of white supremacy is accomplished by assuming the white descendant will inherit the father's dynasty and enriches it more and more. In the same way Bibo decides his descendant's legitimacy and racial category in W.C. Falkner's *The Siege of Monterey*, Sutpen arbitrarily decides that Bon is not his proper son, and chooses Henry as his white legitimate son instead, expecting his descendant's success in the future. Shreve mocks Sutpen's prospect for the future and suggests the arbitrary distinction of race is only stabilized by the narrator's monologic discourse and its unproven expectations for the future. Once other perspectives are introduced to the racial narrative, racial distinctions cannot be stabilized. This novel's structure reveals that fixed racial categories are the result of the exclusion of the dialogical relationships among individuals.

Shreve wants another interpretation of Sutpen's story in order to continue his exciting conversation with Quentin. Shreve's demand to continue storytelling reveals that such cooperative narratives cannot stabilize the order of Sutpen's community. Shreve's narrative reinterprets Sutpen's blood relationship expanding beyond racial borders as the new formation of social connections, which requires the reconsideration of the monologic stability of Sutpen's design and negates the illusion of Sutpen's complete dominance of his family members. In this case, Shreve, who believes in white supremacy and white purity, projects his own white-oriented worldview on Sutpen's drama, but his will to continue the narrative with Quentin inevitably makes the character's racial category unstable.

In contrast to Shreve, Quentin, who expected that his intersubjective narrative with Shreve would stabilize the story of Sutpen, tragically finds their dialogue adds yet another version to their tale. Quentin, who desperately wants to preserve a unified narrative, finally rejects Shreve's request to continue their dialogue. Near the end of their dialogue, Quentin sinks into the monotonous repetition of the same phrase, as if demanding to return to the monologic world. Refusing to answer Shreve's questions, Quentin literally repeats a word monotonously:

> "And so do you know what I think?" Now he [Shreve] did expect an answer, and now he got one:
> "No," Quentin said.
> "Do you want to know what I think?"
> "No," Quentin said. (AA 302)

Shreve desires Quentin's interest in his interpretation; however, Quentin rejects continuing this dialogue because he finds it neither creates a unified worldview nor does it recreate the Southern old glory and homogeneity. In his dialogic space, Quentin finds there is no unified stability of his worldview if he has relationships with others, thinking "Nevermore of peace. Nevermore of peace. Nevermore. Nevermore. Nevermore" (AA 298–99). In contrast, Shreve enjoys telling the story with Quentin. There is no evidence of truth in Shreve's interpretation of the story, but Shreve's exciting narrative shows that dialogic relationships can create a living sensation with others in its multiple possibilities. His interpretation requires sharing his excitement with Quentin, and he wants to continue their unstable narratives to create a new future.

At the conclusion of their dialogue, Shreve asks Quentin, "Why do you hate the South?" (AA 303). Shreve supposes that Quentin hates the monologic South because it prevents plural visions and dialogical relationships. However, contrary to Shreve's expectation, Quentin declares that he loves his land and its monologic worldview, which allows him to keep his obsession of incest and death without any instability. Quentin monotonously repeats, "*I dont. I dont! I dont hate it! I dont hate it!*" (AA 303). The performative effect of Quentin's words is more important than what Quentin means. By monotonously repeating such negative utterances, Quentin suggests a return to the monologic world of Yoknapatawpha and declares his love for it, which might be essentially the same as W.C. Falkner's imagined space where the narrator can create a romantically unified world. W.C. Falkner's romantic world is where Quentin is allowed to express his uncontaminated illusion of incestuous unity between him and Caddy.

Shreve's words suggest modern people hate the American South because of the Southern white elite worldview, which is too self-centered and celebrates the eternal prosperity of their own race. This worldview even boasts homogeneous, incestuous relationships to express their sense of unity. Shreve's dialogue with Quentin reveals that the Southern elites have already consumed their own future by their self-sufficient plan. It is a literary suicide in which they inaugurate their life in their narrative. They have already enjoyed their lives' whole vision or their future prosperity in their own imagination, and they cannot create their new futures by their own hands, actual works, or negotiations with others.

Henry's death in Sutpen's corrupted big house symbolizes how the American South has consumed their future because of its monologic worldview. For Quentin, the fatality of Sutpen's monologism is symbolized by his inheritor, Henry. To find how Sutpen's ambition ended with his legitimate son, Rosa and Quentin visited Sutpen's house in 1909. There Rosa and Quentin discovered Henry had returned to Sutpen's house four years ago only "*to die*" (AA 298). Henry had "the wasted yellow face with closed, almost transparent eyelids on the pillow, the wasted hands crossed on the breast as if he were already a corpse" (AA 298). His appearance and his repetition of the words "*to die*" suggested how a person who gravitated toward the monologic vision of Yoknapatawpha could not find any other place to complete his life. Yoknapatawpha County was only the place where a white male could demand the whole obedience from his family, and Henry's return to Yoknapatawpha meant Henry could not end his life as a decent white Southerner without that obedience and protection. As Henry expected, "Daughter of Thomas Sutpen and a negro slave" (AA 308), Clytie, loyally took care of Henry. Henry's deathbed condition was a sadder version of the Southern landowner whose ideal condition Sutpen once imagined as living "in a barrel stave hammock between two trees, with his shoes off and a nigger . . . who did nothing else but fan him and bring him drinks" (AA 184). As if he was a landowner, Henry demanded the complete obedience of those inside his own family, and this was only possible in Yoknapatawpha. Quentin, who witnessed Henry's deathbed, also understands that the monologic worldview does not result in the prosperity of the descendants, but instead brings ruin and death to the community's members. Remembering that he had compassion with Henry, it is recognized that Quentin has also inherited the fate of Southern monologism, and Henry's deathbed would be the symbolic result of the Southern male's monologic worldview. Near to the end of his conversation with Shreve, Quentin refuses to hear Shreve's words "without listening or answering" (AA 299). Here Quentin quits his dialogic relationship with Shreve and slips into his own monologic fate, conceiving the premonition of his own death. In this concluding part, Faulkner seems to self-referentially analyze the main reason of Quentin's suicide in *The Sound and the Fury*. After he finds his conversation with the world create unwanted new possibilities, Quentin keeps away from Shreve and chooses silence and death to maintain his own obsessive monologic world.

Before Quentin falls into his self-destructive silence, Shreve has dialogue with Quentin and critically examines Sutpen's life to explain why his design does not accomplish the promised prosperity, the acquisition of the ideal inheritor of his dynasty (his white son), or the refutation of his past faults. Principally, Shreve and Quentin guess Sutpen's plan to get "richer and richer" (AA 209) as a planter became impossible during and after the Civil War because of his lack of communication with the others. Sutpen lost his slaves, and though he cooperated with a poor white man, Wash, to start a little store after the war, Sutpen arrogantly refused to communicate with the customers. Based on the narratives of Rosa and his father, Quentin imagines Sutpen "would get mad at his customers, the niggers and the trash and the haggling, and turn them out and lock the door and drink himself blind" (AA 225). It seems that this supposition is the "best of ratiocination" for both of two young narrators (AA 225). However, when Shreve discovers that Sutpen finally failed to get the white legitimate son even after he slept with Wash's granddaughter, he himself wants to explain again why Sutpen lost his first and second sons. Shreve exaggerates Sutpen's inability to communicate with the other, and retrospectively narrates that, near the end of the Civil War era, Bon appeared to have some conversation with his father, but Sutpen sent him "*no word*" and did not recognize him as his son (AA 285). To Quentin's regret, Shreve develops his narrative more dramatically and connect it with Quentin's one. Shreve considers Bon, who had already had his own "octoroon" wife and her child (AA 271), demanded to take revenge on his father and did not become the member of Sutpen's family but became "*the nigger that's going to sleep*" with Judith (AA 286); Henry shot Bon and disappears from Yoknapatawpha—Sutpen lost his first son, and his ideal white inheritor at the same time. After the war, Sutpen slept with Wash's granddaughter in order to have another white son, but Sutpen found it was not his "son" but "a girl" (AA 234). Sutpen totally failed to gain a white male heir and he derided Wash's granddaughter. Wash heard Sutpen's scornful words and killed Sutpen. Sutpen gave no prosperity to his descendants, but caused his and his descendants' ruin because of his arrogant discrimination of the others.

In Shreve's imagination, the socially marginalized people, the alternate story that is separated from Sutpen's monologism, still live in Yoknapatawpha. The story continues. Shreve uses Jim Bond, Charles Bon's grandson who haunts Sutpen's house, as the symbol of the continuation of people's dialogical relationships. As if suggesting that monologic attitude cannot perceive the prosperous effect of a dialogical relationship with others, Shreve says though Quentin cannot see Jim Bond he always exists: "You've got one nigger left. One nigger Sutpen left. Of course you cant catch him and you dont even always see him and you never will be able to use him" (AA 302). If it is possible to interpret Jim Bond as the symbol of dialogic space, Shreve suggests Quentin can see and use him when Quentin want to create another form of society with others.

As Bakhtin suggests, dialogism creates an "open-ended" condition in the world (PDP 298). In spite of Sutpen's conviction of his own design, the dialogue between Shreve and Quentin reveals that its uncertainty continues. One utterance always has a limitation, and it does not arrive at the finalized truth in its conversation with others. There is no final answer for this world, and the uneasy condition goes on. No one can completely guess what will happen in the future. Shreve's imagination has no evidence, but its uncertainty reveals that one utterance never has the power to stabilize destiny. Shreve enjoys his arbitrariness of his narrative, and he is excited to continue his narrative to resist the Southern monologic worldview. Yet it also means that Shreve's narrative is now far away from Quentin's desire to create unity within a two-person, cooperative narrative.

Shreve takes part in the story of the American South in the same way international readers can take part in Faulkner's world and disturb the Southern monologic structure. Shreve's dialogue with Quentin reveals the relativized relationship between narratives and the openness of the story's interpretation to everyone. Anyone can guess the story's truth, and in this process it is inevitable that the story finally becomes part of the narrators' dialogue, in which the participants explore the new formation of their society. Shreve's last comment about the African king shows that the process of revising our social narratives also foresees the new social structure that can contain multiple heterogeneous elements. The alternative future can be inspired from the dialogue, not from a monologic obsession with one design. Dialogism destabilizes the world in order to reopen the society and its past and to find another future. Our dialogic condition puts the world in the process of incessant examinations. If readers only focus on the monologic aspect of Faulkner's work—Sutpen's design—and want to seize the absolute goal in the narration, it means they are mainly reading W.C. Falkner's romantic world in Faulkner's works.

Monologism is a legacy of W.C. Falkner's works, and is the base of Faulkner's imagined Yoknapatawpha County. Thomas Sutpen is a main character whose life is caught by this monologism. Sutpen wants to build his own narrow community under his complete control, and Yoknapatawpha allows Sutpen to exercise his forceful plan to govern his community's members. Faulkner's modern characters in the twentieth century wrestle with this self-centered worldview, and Quentin, who is caught by its attractive effects, surrenders to the monologic worldview in his narrative at the end of the novel. Faulkner uses W.C. Falkner's monologism and its motifs as the community's discursive discourse and uncovers the typical Southern white male's mindset. Faulkner's early and middle works mainly focus on how Southern white characters are caught in the American South's monologic gravity and fail to compose alternate social forms in their dialogues.

W.C. FALKNER'S MONOLOGISM AND FAULKNER'S "WASH"

W.C. Falkner's *The Spanish Heroine* (1851) can be compared to Faulkner's "Wash" (1934) to understand how these two authors use the brave hero's motif. "Wash" was written and published before *Absalom, Absalom!* (1936) and focused on Wash's motivation to kill Sutpen. I mainly use a humiliating experience, that a person is refused entry into a house, to analyze how Faulkner develops and re-accentuates W.C. Falkner's heroic motif in "Wash." W.C. Falkner's *The Spanish Heroine* illustrates what kind of heroic image white Southerners had in their imagination, and how this image influenced their formation of the community.

Belonging to a Narrow Community: *The Spanish Heroine* and "Wash"

The Spanish Heroine, the first novel of W.C. Falkner, shows two features that are shared with his first epic, *The Siege of Monterey*. First, his romantically imagined world has the pyramidal caste system, and at the top sits the Anglo-Saxon hero. Second, the hero's voice (and the author's) has the power to form his own self-centered community. These two features are key factors to establish the hero's social position. W.C. Falkner's novel shows the individual's career advancement in the nineteenth century, with the presupposition that a white hero, Henry, can achieve his goals by violence and form a new community with his own words.

This is a poor American white boy's rags-to-riches story, and his dramatic move from a humiliated position to the honorable citizen comprises the main plot. Henry leaves his house hating his stepfather and starts a life as a poor boy on the road. He leaves his loving mother and goes out into the world "in search of fortune and fame" (SH 3). In the middle of his trip, he meets a wealthy Spanish family's daughter, Ellen. Ellen's father intends to make his daughter marry the decent Spaniard, Gabbez. Henry and Ellen are separated by the violent intervention of Ellen's father and Gabbez. By a twist of fate, Henry becomes a "young Colonel" (SH 97) in the US army during the US-Mexican War, and Ellen also becomes the Mexican army's "young Spanish officer" (SH 98), wearing a man's military dress. Henry becomes "the gallant young warrior" (SH 97) in the field: "In his right hand he held a ponderous saber, the point of which streamed with human gore, and dying Mexicans marked the way the young officer went" (SH 97). Henry and Ellen met at the battle of Monterey as foes, but Ellen saves Henry's life in spite of her position in the field. When Gabbez tries to kill Henry with a lance, Ellen hacks off Gabbez's arm "with the lance still clenched in the severed hand" (SH 108).

Henry and Ellen are then captured by the Mexican army, but they escape from Mexico to New Orleans with the help of Ellen's servant, Pedro. In this process, Pedro meets his decent fiancée in New Mexico. Finally, Henry, who is famous for "his gallant deeds in Monterey" (SH 130), marries Ellen. At the end of this novel, the hero discovers his stepfather died and his loving mother comes to his marriage ceremony to be a member of his family. On the same day, Pedro also marries his wealthy lover and becomes a member of Henry's small community with complete obedience to Henry. In this new community, Henry establishes his and his wife's absolute supremacy. Henry and Pedro become wealthy citizens in Kentucky.

In W.C. Falkner's work, the white hero's social honor and his voice's authority are strongly connected to his power to conquer a particular area. For example, Ellen frequently imagines Henry's gallant massacres of the enemy with his leading shout while dressed in prestigious costume in her dream:

> While she slept her young breast swelled with emotions of delight, for she was dreaming of Henry. In her vision she sees her lover mounted on a noble white war-horse, with a star of honor shining on his manly breast; two golden epaulets rest on his shoulders; a golden plume streams from his military hat, and a golden sword swings by his side. His proud white steed, most gorgeously caparisoned, paws the earth as if proud of the burden which he bears. She now sees him surrounded by fire and smoke, and hears the loud roaring of artillery; she sees two mighty armies drawn up in battle array, while her lover, mounted on his white charger, leads on his regiment to the charge amidst shouts of victory, which follow him through the reaking [*sic*] ranks of the enemy. The blood stained banner floats majestically over the head of her lover as he cheers his men on to victory. (SH 62–63)

In her dream, Ellen completes the heroic image of Henry who fights brilliantly and accomplishes the victory. Her dream shows a typical romantic imagination of a heroic soldier on the battlefield, and this image is realized when Henry actually becomes a member of the army of "the Anglo-saxons [*sic*]" who "march along the field" (SH 96) as the "young heroic Colonel" (SH 98). Henry's violent actions on the field are predicted in Ellen's dream, and he later on executes them. This agreement of imagination and reality enforces this romantic story's monologic effect, and the realization of a violent soldier's image is strongly connected with the Anglo-Saxon's power to conquer the enemy. In *The Spanish Heroine*, the story focuses on how the romantic imagination merges with reality to confirm the hero's self-centered order.

In this monologic world, a brave hero's voice actually creates his own pyramidal structure in his narrow community. This class structure is shown in a minor character, Pedro, who is a servant of Ellen's family. The author narrates he was young, "tall, well built, and extremely handsome; his glossy hair was as black as the raven's wing, his skin a little dark, broad shoulders, and full chest; his eyes were a deep blue, and he was as brave as Julius Cæsar" (SH 104). However, Pedro appeared as a character whose position was decided by economic condition. Pedro belonged to a slave class in Mexico, Peons, which consisted of contracted slaves based on the reason of economy. The author explains that this system is different from slavery in the United States:

> Reader, you must not suppose Pedro to be like the servants we have in the United States. There are a certain class of people in Mexico called *Peons*, who are compelled to serve as slaves on account of debt, whose owners always manage to keep them in debt, being allowed by law to charge them for their clothes and board; consequently, instead of growing smaller, the debt always glows larger; and the offspring is always bound for their ancestor's debts, consequently, they are made slaves from one generation to another. (SH 104)

W.C. Falkner explains that the Peons system begins with a purely economic reason, but it is maintained by a cunningly schemed debt. Pedro seems to have the possibility for social mobility, and yet the author presupposes that in any generation and in any size, when people gather, the social hierarchy is reproduced inside their community to stabilize its flexibility.

The Spanish Heroine depicts how the dominant individual creates the social stratification inside his community with their words and minor orders. For example, the class conscious in Cuba is practically shown in a high-class citizen's attitude toward Pedro. Pedro, who helps Ellen meet Henry, is hated by Ellen's father, General Aakenza. Aakenza sells Pedro to a large cotton field in Cuba owned by Aakenza's brother. But soon Ellen purchases Pedro back in Cuba and procures him "a clerkship in a large mercantile establishment" (SH 103). After becoming a wealthy partner of the business, Pedro falls in love with Julia, the only daughter of "Count De Lorenzo, who is one of the wealthiest and most distinguished citizens of Cuba" (SH 103). Lorenzo, who learns that his daughter has fallen in love with Pedro, tells Pedro "if he ever put his foot in his door again his life should pay the forfeit" (SH 104). With this order Lorenzo creates a segregating barrier inside of his personal community. After this separation, Julia's father forcefully sends Julia to England.

In W.C. Falkner's work, characters show some possibilities to create social combinations when they are in incessant mobilization, but once the people settle again to form a society the dominant person's absolute power is resumed to decide the order. In the latter part of this novel, Pedro becomes a wealthy merchant and finds he is the only heir of Count De Gazarro. He becomes an eligible husband for Julia, and a member of bourgeois family society. Though the order of the people is rearranged, the hero's voice retains its dominance.

Henry's voice has the power to make his own order around him. The power of the hero's voice is obvious in the marriage scene of hero and heroine. Until the marriage ceremony of Henry and Ellen begins, Pedro cannot find his separated lover, Julia, and Pedro mournfully wears a black suit. Pedro's heart is broken in his desperate search for Julia, but Henry orders him to cooperate with their wedding. Henry and Ellen command Pedro to wait on him as Henry's groomsman, and to meet Ellen's mysterious bridesmaid:

> "Pedro, you must wait on me," quoth Henry.
>
> "I'll do so, Henry, if you desire it, but I am not acquainted with any lady in the room; and you know that I desire not the acquaintance of any; as I have lost Julia forever, I have no desire to form new acquaintances."
>
> "Nonsense, my dear Pedro, you must strive to forget Julia; perhaps she has forgotten you. She may be at this very moment dancing at a ball in London, and thinking no more about you."
>
> "Stop, Henry, say no more on this subject, if you please; I'll wait on you if you desire it."
>
> "Yes, Pedro, said Ellen," [*sic*] springing forward, "you are to wait with Miss Lizzeal. She is a charming girl, and you will say so when you see her. She looks as much like your Julia as if she was her sister, at least I thought so, as I only saw her face through her veil; but what puzzles me she is dressed in black, and requested that she may wait on me in that garb."
>
> "Well, that'll suit precisely, as I am dressed in black too." (SH 131)

After this conversation, as readers can easily guess, Miss Lizzeal turns out to be Julia, who has disguised herself to veer away from her father's order and search for Pedro. Concurrently, Julia is accompanied by a wealthy countess, who is Pedro's mother. The countess lost Pedro when he was very young and has searched for him for a long time. Now Pedro becomes the son of the countess and a member of the aristocrats. This new economic and familial position suggests a drastic change in the relationship between Pedro and Henry, but Henry still keeps his emotional supremacy over Pedro. In this denouement scene Henry also happily finds that his mother, who had lost her second husband and seemed to have died, comes to attend Henry's marriage. Being into ecstasies over his family reunion, Henry has a rapturous conversation with Pedro:

> "Run here, Pedro," said Henry, "I have found my mother, over whose grave I have shed many a tear."
>
> "Well, if you have found your mother, you have nothing to brag of. You have found your mother, but I have found my mother and Julia too; what do you think of that?"
>
> "Well, well," said Henry, with a smile of levity on his cheek, "if the dead ain't alive and the lost found now, I think it would be good policy to give up the chase." (SH 133–34)

It is astonishing to read Henry's selfish attitude to Pedro. The social order has already been established with the hero's voice and never been altered: Henry never cares about Pedro's warmth of feeling and Pedro endures it. Pedro can enter into Henry's world only by accepting his arrogant supremacy. Or, in other words, W.C. Falkner's monologism cannot describe the change of Henry and Pedro's social relationship in their dialogic condition. In the marriage scene, the order is kept: "Henry and Ellen led off first. . . . Then Pedro led Julia forward" (SH 136). This scene suggests that in their narrow community, Henry and Ellen keep their priority, even when this novel concludes, "Henry and Pedro live in the same beautiful valley, in the chivalric State of old Kentucky, where they have magnificent buildings, beauteous flower gardens" (SH 136). In W.C. Falkner's work, the hero's (and the author's) voice has the power to keep the social order from beginning to end.

As the happy ending of this novel, the hero establishes his dominant power in his narrow community. Around Henry, Pedro and the five women—Julia, Ellen, Pedro's countess mother, Henry's mother, and Ellen's mother—gather to live together. Henry's stepfather has passed away, and Ellen has escaped from her authoritative father, Aakenza. The hindrances to the hero's happiness are eliminated, and the author declares that "there never was a happier circle seen beneath the sky than they—those two young men were Henry and Pedro" (SH 136). Apparently the author seems to celebrate the social combination of Henry and Pedro in this passage, but the whole story suggests Henry's happiness comes from Pedro's obedience. In *The Spanish Heroine*, an ambitious young man rises as a young soldier in the war; he becomes a famous soldier in the United States, and at last he and his heroine are married. Under the dominance of the couple that is celebrated by the author's voice, the differences of gender, nation, and race in each character are settled in a hierarchical order. The hero's absolute power comes from his ability to rearrange the social order arbitrarily. The author allows the hero have racial multiplicity inside of the narrow community, and the hero's monological voice totally controls the other characters' hierarchical arrangements and retains its power to establish Anglo-Saxon dominance.

This arrogant power of the hero's voice is similar to some self-centered characters' voices in Faulkner's works. A typical example might be Thomas Sutpen's voice in "Wash," which demands to create absolute order in the same way Henry does in *The Spanish Heroine*. The difference is that Faulkner reveals how Sutpen's voice fails to create the caste order around him because the other socially marginalized people rebel against Sutpen's dominance.

A Protagonist's Rebellion against Sutpen's Monologic Worldview in "Wash"

Here I take Faulkner's short story, "Wash," as a masterpiece that focuses on the powerless, poor white's demand to be a member of Sutpen's narrow community. Comparing *The Spanish Heroine* with "Wash" is useful in explaining how Sutpen controls Wash. In this short story, Sutpen is described as a person whose voice still retains the authority that is almost the same one that Henry's voice has in his relationship with Pedro. In contrast to Henry who retains his absolute power until the end of the novel, Faulkner describes Sutpen's postbellum condition as the place where marginalized characters furiously resist the protagonist's dominant pressure.

"Wash" depicts that, before the Civil War, Sutpen's plantation plan and his words worked well to certify the order of the society. Sutpen's monologic worldview promises his plantation's eternal prosperity, and Wash—a poor white who lives in a corner of the Sutpen's field—is attracted to Sutpen's worldview. Wash identifies Sutpen's image with God. For Wash, Sutpen was a "fine proud man" and he imagines, "If God Himself was to come down and ride the natural earth, that's what He would aim to look like" (WA 538). The structure of Wash's image of Sutpen is the same as Ellen's dream of Henry; the difference is that Henry realizes Ellen's dream while Sutpen betrays Wash's. Faulkner uses this re-accentuation of the hero's romantic image as the central theme of his tragic drama in which one's monologic voice cannot realize its ideal image.

Wash expects that Sutpen feels some fraternity with him because they share the same skin color. However, this expectation is rejected by Sutpen's personal decision for his narrow community's social structure. While Sutpen was a soldier in the Civil War, his slave daughter, Clytie, prevented Wash from entering Sutpen's house. This scene is more impressively written in "Wash" than *Absalom, Absalom!*. Clytie refused Wash entry by saying, "Stop right dar, white man. Stop right whar you is. You ain't never crossed dese steps whilst Cunnel here, and you ain't ghy' do hit now" (WA 537). Clytie exercised Sutpen's personal sense of segregation as his agent. This relationship between Clytie and Wash is the reverse of the racial relationship between Lorenzo and Pedro in *The Spanish Heroine*—Lorenzo doesn't allow Pedro to "put his foot in his door" (SH 104). These two scenes' racial relationships are different, but both tales show that the social hierarchies are decided by the narrow community's authoritative voice.

Nevertheless, Wash knows Sutpen can easily change the social structure with just a few words. Therefore, after this repudiation from Sutpen's house, Wash consoles himself by believing Sutpen will accept him as a community member in the future. Faulkner describes what Wash has in his mind:

> There was this of a kind of pride: he had never tried to enter the big house, even though he believed that if he had, Sutpen would have received him, permitted him. "But I ain't going to give no black nigger the chance to tell me I can't go nowhere," he said to himself. "I ain't even going to give Kernel the chance to have to cuss a nigger on my account." (WA 537–38)

Wash believes Colonel Sutpen has a fraternal feeling toward Wash. Expecting that Sutpen himself would have allowed Wash to enter his house if he wanted, Wash did not try Sutpen's will because of his trust in Sutpen. He imagines that, in the same way Henry finally makes a cordial relationship with Pedro as long as Pedro faithfully obeys Henry's order in *The Spanish Heroine*, he could be a member of Sutpen's intimate community with his honest loyalty to Sutpen.

Sutpen's family condition suggests that the white male's narrow community in Yoknapatawpha contains an interracial element. Sutpen's absolute dominance keeps Clytie, his slaveborn daughter, under his control without disturbing their class consciousness, and makes her reject Wash's entering his house. Apparently, the racial category works as the main factor to decide the social hierarchy, but "Wash" and *The Spanish Heroine* tell readers that the internalization of a hero's imagined social order is a key element to prove that the other social members' belongs to the hero's world. Wash expects Sutpen's monologism will finally allow him to merge his own ideal world with Sutpen's. Wash assumes he and Sutpen are in the same category because he has already internalized Sutpen's imagined social order. This is how Wash belongs to Sutpen's world.

In his imagination, Wash believes he belongs to the same category as Sutpen. Therefore, Wash does not lose his hope when he witnesses Sutpen's impecuniosity after the Civil War. Sutpen loses his slaves and buys a small shop. As if to conceal his powerlessness, Sutpen never ceases believing in his dominance within his own narrow community. In this condition, Wash and Sutpen seem to have the opportunity to have an equal relationship. In the little shop, Wash and Sutpen drink inferior whisky "now together from a stoneware jug" (WA 539). However, in their conversation, Sutpen's monologic worldview, which is full of Sutpen's stubborn will to realize his own future design without any consideration for another consciousness with equal rights and equal responsibilities, still has its dominance: "the talk would not be quiet now, as when Sutpen lay in the hammock, delivering an arrogant monologue while Wash squatted guffawing against his post" (WA 539). Even in this miserable reconstruction days, Sutpen's words continue his monologic worldview and Wash hears it with delusive belief.

Even in the poor condition after the Civil War, the relation of Sutpen and Wash is pervasively unchangeable. After they drink together "in the rear of the little store" (WA 539), Wash carries Sutpen into his house: "He entered the house now. He had been doing so for a long time, taking Sutpen home in whatever borrowed wagon might be, talking him into locomotion with cajoling murmurs as though he were a horse, a stallion himself. The daughter would meet them and hold open the door without a word" (WA 540). Wash can now enter his house, but Wash still believes his position is lower than Sutpen. Putting Sutpen in bed, "he would lie down on the floor beside the bed" (WA 540). Wash believes in Sutpen's superiority, and his belonging means the acceptance of his own inferior position inside Sutpen's community. In Sutpen's monologic world, the people who enter Sutpen's place should accept they are subject to the order of Sutpen's dominance.

What Wash believes is a simple story. His hero is an Anglo-Saxon elite, a noble race who can be brave in the battlefield and gain fortune and decent descendants. Wash believes these images and demonstrates that social class and segregation are, in Sutpen's world, not reproduced from money and race but from this monologic story Sutpen's community members share. In Yoknapatawpha's romantic, monologic gravity, the white hero decides his narrow community's order following this illusion. Wash believes that the Anglo-Saxon hero will "set the order and the rule of living" (WA 547). In Wash's imagination, the Anglo-Saxon hero's will is the main principle of everything.

When it is discovered Sutpen seduced Wash's granddaughter, Wash develops a family image the romanticized heroic story suggests, believing it is honorable for his granddaughter to be a hero's wife, even if the hero is almost the same age as Wash. In the following scene, Sutpen asks Wash if he is afraid of Sutpen, mentioning his unscrupulous seduction of Wash's young granddaughter, but Wash says he is not afraid of his granddaughter's physical relationship with Sutpen because Sutpen is a chosen man. For Wash, Sutpen is still a brave hero who will prosper in the future:

> Now Wash's gaze no longer questioned. It was tranquil, serene. "I ain't afraid. Because you air brave. It ain't that you were a brave man at one minute or day of your life and got a paper to show hit from General Lee. But you air brave, the same as you air alive and breathing. That's where hit's different. Hit don't need no ticket from nobody to tell me that. And I know that whatever you handle or tech, whether hit's a regiment of men or a ignorant gal or just a hound dog, that you will make hit right."
>
> Now it was Sutpen who looked away, turning suddenly, brusquely. "Get the jug," he said sharply.
>
> "Sho, Kernel," Wash said. (WA 541–42)

Readers can guess that Sutpen looks away because he knows he is not an ideal hero. This short story's whole structure reveals that Sutpen is not a typical heroic figure. Maybe he did not attain the brilliant honor on the battlefield as expected. He knows that he is losing power to realize his ideal world. Moreover, the powerlessness of Sutpen's words is revealed when Sutpen fails to have a white son with Wash's granddaughter. Wash realizes Sutpen's monologism does not have the power to realize the future as he said. This recognition demolishes Wash's illusion that Sutpen has the ability to fulfill his heroic vision, or to "make hit right" (WA 542). Ironically, when Wash realized "Sutpen was an old man like himself" (WA 543), Wash's own pride as a member of Sutpen's community was also demolished.

Wash believed Sutpen had the power to control the destiny of his community members, but, when Wash's granddaughter had a girl, Wash was betrayed because Sutpen's arbitrary order could not create the society Sutpen desired. Wash was so obsessed with the mythic dream that he really hoped eternal bliss would come when Sutpen had a legitimate son again. However, Wash discovers it was an illusion. He loses both his confidence in Sutpen and his hope for the future. All Sutpen really has is his arbitrary will to decide the insiders and outsiders of his narrow community, and Sutpen finally decides that Wash and his family members are not insiders. Because of this crucial decision, Wash desperately kills Sutpen for revenge. Wash may not have wanted an equal relationship with Sutpen, but he finds there is no reason to live in a world where the hero has lost the dominant power but still rejects the Washes from his narrow community.

From the start of the story, Wash willingly shares his illusion with Sutpen, and never dreams of sharing an equal position with Sutpen. This condition is the same as Pedro's in *The Spanish Heroine*. Pedro chooses to be the subordinate member in his relationship with Henry and Ellen, and he gains a peaceful position in the narrow community as well as prosperity. Because of this mutual agreement, at the end of *The Spanish Heroine*, Henry shows his power to create his own narrow community in Kentucky by including Pedro as a member of his community under the proper order. In Wash's case, Wash wanted to make sure of his position with Sutpen using his granddaughter's marriage, but Sutpen fails to create the new community's order, which should have included Wash, his granddaughter, and their newborn baby. Wash finds Sutpen's power to create an ideal world is fake when Sutpen fails to have white son and finds no longer values Wash's granddaughter.

In his words, Sutpen is cruel enough to betray that he does not want to form a new social order with Wash, Wash's granddaughter, and her child. Instead, Sutpen prefers to make a new stable for his mare, who gives him a good colt. When Wash hears Sutpen tells his granddaughter that if she was a mare he "could give her a good stall in the stable," Wash feels "something seemed to stop dead in him" (WA 544). This "something" is Wash's illusion of a small community with Sutpen at the head. Wash believes in this monologic illusion of Sutpen's future prosperity and his family's belonging to this community. When this dream is corrupted, Wash's future is also demolished. Wash kills Sutpen with a scythe he borrowed from Sutpen, symbolizing that what motivated Wash to live was his borrowed monologism from Sutpen, and this failed ideal motivates him to kill Sutpen. He also kills his granddaughter and her baby by himself and then rushes at the sheriff intentionally to be shot. Once he loses his illusion of future prosperity with Sutpen, he does not want him and his relatives to live in such a desperate world.

Sutpen is not strictly following the so-called "one-drop rule" to decide the social hierarchy. Rather, only his arbitrary will decides his narrow community's order. Sutpen made a negro slave bear his daughter, Clytie, and he lets her live inside his house. Clytie loyally protects Sutpen's house and prevents the poor white, Wash, from entering. Sutpen's monologism dominates the people's class conscious inside his narrow community, and his power is exercised when he decides who can be in his house and who cannot.

Monologism in the imagined world is typically realized in W.C. Falkner's *The Spanish Heroine*. In his first novel, W.C. Falkner insists the hero's happiness lies in his ability to make a narrow community under his dominance and to achieve eternal prosperity. The author identifies the hero's successful achievement as his one and almost at the end of the novel yells out: "Now, reader, imagine just what you d—n please, for I am going to throw down my pen shortly, and shout for joy; for I can't describe any more happiness without shouting some myself" (SH 136). The rapturous author cerebrates this union without any limitation. This ending suggests W.C. Falkner believes a narrow community under the dominance of a white male, who can be identified with the author, is the happiest community in the world, and he crystalizes it in a simple, discursive vision. This tale also suggests what kind of typical goal is supposed in the main characters' imagination in "Wash."

Blotner feels *The Spanish Heroine* is "wildly improbable coincidences and amateurish wrenching of situations to meet the demands of a mechanical plot made this second effort decidedly less accomplished than the first."[88] However, this bad novel reveals what kind of romantic pattern W.C. Falkner conceives in his imagination. In his typical pattern, the protagonist creates the social hierarchy by himself by expelling foes, and he arbitrarily chooses his community members, and even family members, to be under his dominance. Faulkner's modern characters inherit this pattern in their shared imagination. What Wash believes is the hero's power to control this "improbable coincidences." W.C. Falkner's first novel can be considered the literary demonstration of the Southerners' typical belief in monologism.

In "Wash," Faulkner re-accentuates W.C. Falkner's monologic pattern and reveals how the hero's worldview overlooks the possibilities of egalitarian relationships. Faulkner depicts Sutpen's arrogant aspect from Wash's point of view, and illustrates how his selfish design is revenged by the socially marginalized character's frustrated emotion. Sutpen imitates a typical heroic worldview to foretell his success in the future, but his real life fails to attain this prosperity. Wash, who shares his imagination with Sutpen, expects that Sutpen's improbable power will make his own design real and therefore expects his granddaughter to have a son, following Sutpen's design. Wash's murder suggests Wash finally recognizes Sutpen's illusion does not have the power to dominate even his narrow society. Wash is a tragic person who believed in Sutpen's ambition and could not find any other social forms for himself when that illusion was shattered. Losing his monologic model, Wash desperately kills Sutpen, his descendants, and himself. In the next section, I compare W.C. Falkner's *White Rose of Memphis* with Faulkner's *Flags in the Dust* to illustrate how Faulkner criticizes his great-grandfather's ideal postmortem image in his work.

THE DECENT FAMILIES IN HEAVEN: *THE WHITE ROSE OF MEMPHIS* AND *FLAGS IN THE DUST*

W.C. Falkner's romantic pattern of the family union demands the eternity of the white male's will to dominate the members of his narrow community even after his death. This desire is expressed in a sophisticated way in *The White Rose of Memphis*. At the end of this bestseller novel, W.C. Falkner depicts his two ideal white couples in everlasting glory. Three years after they finish their trip on the luxurious ship named *The White Rose of Memphis*, Edward Demar and Harry Wallingford, who are happy with their wealthy wives, Lottie and Viola, have descendants. They typify rich families in the American South: Lottie and Viola become "two elegantly dressed ladies" who "were leisurely strolling along the graveled walks in Court Square, Memphis, Tennessee, engaged in an earnest conversation; while two mulatto girls were pushing a couple of silver-mounted baby carriages along just behind the two ladies" (WRM 537). The eternal bliss of these two families of Edward and Harry seem to be guaranteed by their beautiful white children. The narrator praises Harry's "bright, blue-eyed boy" who has "bright, brown curly hair" and describes fathers and their children idyllically: "Doctor Demar lifted his pretty little daughter from the ground as she came running to meet him, and seated her on his shoulder and began to dance around with her, while Harry began to romp with his boy, who was galloping about astraddle of a stick which he was lashing with a whip" (WRM 541-42). These white families' happy conditions are celebrated by the community's eldest member, Nathaniel Rockland, a lawyer in Memphis who raised and supported Harry and Lottie as their uncle. At the end of this novel, he appears as the patriarch of the families who enjoys their happy unions. He expects that these family unions will continue in heaven after his death, illustrating how the white elite characters in W.C. Falkner's literature imagine their family union's celestial eternity:

> Harry Wallingford purchased a handsome residence, situated in the midst of a beautiful grove of native poplars, just east of Mr. Rockland's, and not more than two hundred yards from it. Viola had her flower garden laid off, so as to adjoin Lottie's, and they were only separated by a clean gravel walk. Those two devoted friends would often seat themselves on a rustic bench in the garden, spending long hours watching the two little children as they frolicked like young lambs on the green turf of the lawn.
>
> . . .
>
> Mr. Rockland, whose hair had grown very white, was sitting on an easy chair on the front portico, watching the happy group on the green lawn.
>
> "I thank God," he muttered, "because He was so good as to send Lottie Wallingford to direct my steps toward Heaven, and to be my guardian angel on earth." (WRM 541–42)

The main characters' peaceful condition in this novel is more sophisticated than the hero's fanciful prosperity in *The Spanish Heroine*. The eldest character realistically extends his expectation more by demanding the divine celebration continue in heaven. For Rockland, Lottie can be his "guardian angel on earth" to direct him from his earthly life toward heaven. This happy-ever-after ending proves this novel belongs to the Southern romantic author's fairly-tale tradition. They are confident their happy life will continue even after their death when they witness that their strong connection has been realized on the earth.

In his introduction, Cantwell evaluates *The White Rose of Memphis* as a local novel that is full of "a sense of detail" and expresses "concrete Tennessee realities of roads, houses, farmers, dogs, weather, narrow escapes, and over the whole adventure a sense of premature responsibility and premature release."[89] Cantwell appreciates this novel's local elements, intending to put W.C. Falkner's *The White Rose of Memphis* as an ascendant of Faulkner's modern regional novels. But saying W.C. Falkner's imagination is local is not so appropriate when we consider that W.C. Falkner's other works tell of a very international issue—the US-Mexican War in both *The Siege of Monterey* and *The Spanish Heroine*. It might be more appropriate to consider that W.C. Falkner extends his monologism to his own region to describe this area's planation culture as an ideal one that is quite near to heaven. W.C. Falkner's imagination is not limited to the American South; rather, he uses the American South as an example of a place where the white race's desire for dominance of their narrow communities is idealistically realized.

W.C. Falkner's great-grandson, Faulkner, criticizes this monologic dominance in this region and describes how this illusion still lurks in the minds of Yoknapatawpha's white males, who believe their honorable position will continue eternally and be promised in heaven.

In *The White Rose of Memphis*, W.C. Falkner declares that the formation of a small, family-oriented community under the domination of an Anglo-Saxon male is an ideal unit whose eternal glory continues after death. This motif is more crystalized and re-accentuated in Faulkner's first Yoknapatawpha novel, *Flags in the Dust*. It has been argued that W.C. Falkner is a model of Colonel John Sartoris in this novel.[90] However, more precisely, Colonel John Sartoris's posthumous vision takes after Nathaniel Rockland's one in *The White Rose of Memphis*. In the same way Rockland oversees his relatives and expects his life will continue in heaven, the statue of Colonel John Sartoris

> stood on a stone pedestal, in his frock coat and bare-headed, one leg slightly advanced and one hand resting lightly on the stone pylon beside him. His head was lifted a little in that gesture of haughty pride which repeated itself generation after generation with a fateful fidelity, his back to the world and his carven eyes gazing out across the valley where his railroad ran and the blue changeless hills beyond, and beyond that, the ramparts of infinity itself. (FD 870)

Though Faulkner's tone is sarcastic here, he modeled his great-grandfather's imagined character in *The White Rose of Memphis* to create this fictional statue's gaze at the eternal world. In the same way as Rockland sees his and his relatives' happy union as a promise to guide his steps "toward Heaven" (WRM 542), the statue of Colonel John Sartoris looks out across "the valley where his railroad ran" and gazes at "infinity itself" (FD 870). These two novels share the same imagined structure: the patriarchal character looks for the eternal celebration in heaven beyond his dominance of his relatives and his land. Because of this imagined structure, W.C. Falkner establishes his own carved image beside his grave and Faulkner models it as his fictional character's statue. In both cases, the imagination and expectations for the eternal prosperity appear as the center of the order. The imagination comes first, and the facts follow it. It is not true that W.C. Falkner's real marble figure causes Faulkner's fictional description of Colonel John Sartoris's sculpture; rather, because of the effect of the shared imagination of the Fa(u)lkners', W.C. Falkner's real statue is settled and it can be a model for the statue of Colonel John Sartoris, a symbol of the fictional family's obsessive expectation for eternity. Faulkner re-accentuates this shared eternal vision as the imagined gravity in *Flags in the Dust*.

Colonel John Sartoris's statue indicates the family's glory after his death, and the Sartoris men inherit this illusion of the brave hero and expect glory in heaven after their violent actions on the earth. It is criticized by Miss Jenny as "a game outmoded and played with pawns shaped too late and to an old dead pattern" (FD 875). However, this old pattern of the male members' heroic imagination has been enforced by the statue of John Sartoris and overshadows the female family member's bitter comment. Because of this sympathetic resonance between the imagined world and the real, being a Sartoris has "a glamorous fatality" (FD 875) that promises a honorable position in heaven. Miss Jenny's voice is sharp but weak in this novel, because even if she criticizes the arrogant attitude of the Sartoris male members, her words suggest that she is also involved in this fantastic worldview when she comments they are God's "pawns" and "the Player Himself is a little wearied" (FD 875). Even if God is wearied, Miss Jenny believes that God still watches the Sartoris family. Because the continuation of the relationship between their family and God, the Sartoris family's male members will not cease their violent actions, trusting in their proper positions in heaven. This lingering belief is expected to be moderated by the new member of the family. Marriage brings Benbow family's blood to the Sartoris family, and Narcissa names her son "Benbow Sartoris" (FD 874) to soften Sartoris's temper with the more peaceful temper of the Benbows, though, as Miss Jenny criticizes, even such naming cannot control the Sartoris tradition.

Almost all of the previous studies that explore the relationship of both Fa(u)lkners' works insist *The White Rose of Memphis* is W.C. Falkner's central work. However, their focus on this bestselling novel has made the other relationships of the two Fa(u)lkners' works more vague. Moreover, there are not as many common motifs between *The White Rose of Memphis* and Faulkner's works. Because of their focus on this work alone, many scholars suppose that there are not strong connections between these two authors' imaginations. *The White Rose of Memphis* should be analyzed as simply one example in the development of W.C. Falkner's monological works to explore the relationships between W.C. Falkner's works and Faulkner's.

An important vision of W.C. Falkner's imagination is revealed at the end of this novel. In *The White Rose of Memphis*, W.C. Falkner expands his monologic pattern to make the patriarchal figure's glory continue after death. Faulkner parodies this pattern and depicts how the Sartoris family in Yoknapatawpha is haunted by this romantic worldview. Next, in the comparison between W.C. Falkner's *The Little Brick Church* and Faulkner's *Intruder in the Dust*, I prove how Faulkner traces W.C. Falkner's monologic imagination in Gavin Stevens's modern monologue.

FROM MONOLOGISM TO DIALOGISM: *THE LITTLE BRICK CHURCH* AND *INTRUDER IN THE DUST*

Faulkner's characters in his early and middle works, especially his white elite ones, are attracted to W.C. Falkner's monologic perspective and identify themselves with the hero's self-contented world, as if W.C. Falkner's romantic discourse were the gravity of Faulkner's mythic world. Faulkner's main characters in his early and middle works try to resist this monologic gravity in their isolated agonies in vain; however, in his later works—which begin with *Intruder in the Dust*—Faulkner focuses on character's social combinations to depict the ways to resist this tempting force. In the following argument, I compare W.C. Falkner's *The Little Brick Church* with Faulkner's *Intruder in the Dust* to show how *The Little Brick Church* insists on the Southern plantation culture's connection with the origin of America, and how this connection naturalizes Gavin's paternal attitude to the African Americans, women and young people in *Intruder in the Dust*. I also illustrate that this structure is criticized in the young protagonist's negative attitude toward Gavin's arguments.

The Hero's Will to Protect the Weak and Helpless in *The Little Brick Church*

W.C. Falkner's *The Little Brick Church* might be less popular and less mentioned than his previous bestselling novel, *The White Rose of Memphis*, but this novel is still beneficial because it shows how W.C. Falkner uses the idea of liberty to effectively propagate that the American South's plantation culture based on slavery is a true inheritor of the landowner's generous spirit in Colonial America. This novel centers on a colonial landowner family, the Falklands, as an archetype of the plantation in the American South. The hero of this family is Oscar Falkland, a brave soldier with a liberal spirit during the American Revolutionary War. The Falklands live in the Hudson River area and seem to have no relationship with the American South, but W.C. Falkner's rhetoric demonstrates how the landowner family in Colonial America can be found as the origin of Southern plantation culture.

W.C. Falkner's ideological intention is to connect the origin of America with the plantation culture of the American South—a broadly shared attitude among the popular Southern authors in the late nineteenth and early twentieth century. Jeremy Wells argues in *Romances of the White Man's Burden* (2011) that it is too much to say that late nineteenth- and early twentieth-century Southern novelists imagined the South as "the true center of 'America,'" but Southern authors Thomas Dixon Jr. and Thomas Nelson Page created "the narrative of origins that made white southern men seem directly descended from the nation's Founding Fathers" and conceptualized "the plantation as the model for a new American imperium."[91] A typical example is Thomas Dixon Jr.'s *The Leopard's Spots: A Romance of the White Man's Burden, 1865–1900* (1902). Wells argues how Dixon creates "linkages between plantation nostalgia, U.S. imperialism, and the new nationalism emergent during these decades"[92] in this novel:

> *The Leopard's Spots* begins with the end of the Old South, ends with the old regime's virtual restoration at the turn of the century, and in its middle chapters manages to represent the rise of the Ku Klux Klan, the overthrow of the Reconstruction governments, the disenfranchisement of "the negro," and the dawn of the age of U.S. overseas imperialism as interlinked episodes in the long and triumphal history of "Anglo-Saxon civilisation."[93]

Dixon's ideological intent is the same as his other notorious novel, *The Clansman: An Historical Romance of the Ku Klux Klan* (1905). In his imagined world, the antebellum South is a paradise both for the black and the white because the white planters are the ideal guardians of the ignorant slaves. Enfranchising black men is not a good idea because the white elites—who are the rightful inheritors of Anglo-Saxon culture—are the ideal leaders to dominant the nation, and even other nations. As Wells concludes, the Southern writer's imagined South is "the conceptual territories" that still generate the American notion of race, region, and empire.[94]

What Dixon shares with W.C. Falkner is the monologic narrative. In Dixon's novel, *The Leopard's Spots*, the author and the narrator are both identified in its monologic discourse, which directly connects the racial problem of the local area to the crisis of the future of the nation. In the following scene, the declared voice is monologically identified with the author's:

> The speaking began at nine o'clock. Every discordant element of the old South's furious political passions was now melted into harmonious unity. Whig and Democrat who had fought one another with relentless hatred sat side by side on that platform. Secessionist and Unionist now clasped hands. It was a White Man's Party, and against it stood in solid array the Black Man's Party, led by Simon Legree.
>
> Henceforth there could be but one issue, are you a White Man or a Negro?
>
> They declared there was but one question to be settled: —
>
> "*Shall the future American be an Anglo-Saxon or a Mulatto*?"
>
> These determined impassioned men believed that this question was more important than any theory of tariff or finance and that it was larger than the South, or even the nation, and held in its solution the brightest hopes of the progress of the human race. And they believed that they were ordained of God in this crisis to give this question its first authoritative answer.[95]

The narrator's ideological speech monologically reflects the author's intention and the orators declare the American South is the central place to decide "*the future America*." As Wells argues, *The Leopard's Spots* connects the monologic narrative of the American South's story with the future of the nation's Anglo-Saxon dominance.

In *The Little Brick Church*, W.C. Falkner uses monologism to identify the Southern culture with the ideal landowner's life in Colonial America. This novel's centenarian narrator, Braddock Barnard, connects the Northern landowner's colonial farm life and the Southern plantation's rural life, as if these two social structures represented America's ideal condition. Barnard visited the Rockland farm when he was seven years old in 1757. He witnessed the Rockland farm's Falkland family, which used slaves and hired hands in the American Revolution era, and believes it is the "perfect system" (LBC 61). He confesses: "I have known no family circle where a greater degree of genuine happiness prevailed than that which blessed the Falkland household" (LBC 61). Later on, Barnard remembers the Southern rural life he witnessed when he visited the South in the early nineteenth century:

> In 1836 and 1837 I travelled over many of the Southern States on business for a New York land company. . . . I never witnessed an instance of wanton cruelty on the part of a master. The negroes were apparently happy and contented, well fed and clad, and when sick the best medical attention was employed for them. The large planters always built comfortable cabins for their slaves. (LBC 114)

Bernard concludes that "the negroes were the happiest race of people I ever saw" (LBC 114) and criticizes Northern capitalism, which cruelly exploits its laborers, "pale-faced little girls" (LBC 127), suggesting the Anglo-Saxon landowner's life who harmoniously dominate slaves and hired hands before the American Revolution is an ideal social form that can be on the same wavelength as the Southern plantation family and its use of slavery.

To naturally connect the budding social condition of Colonial America with the plantation system of the American South, the narrator of *The Little Brick Church* believes American liberty means the will to protect the weak and helpless from the enemy. Barnard, who took part in the American Revolutionary War, argues "the liberty for which I fought was that which I fondly hoped would shield the weak from the oppressions of the strong" (LBC 67). The white elites' liberalism, which rightfully inherits the Anglo-Saxon's historical legacy, is properly ordained as the regnant spirit of the world. This is the theme of *The Little Brick Church,* which sees Colonial America's romantic tale as the origin of plantation culture in the American South. This theme is supported by the narrator's rhetoric. In the following argument, I outline three tactics W.C. Falkner uses to connect the origin of America with

Southern plantation culture. In short, W.C. Falkner's romantic works insist that their white protagonists inevitably demand to protect and govern the other social members when they follow their own natural instincts. I examine how W.C. Falkner ideologically symbolizes the hero's imperial impulse as the core of America's budding liberal spirit in this novel, and how this theme naturalizes the white male protagonist's arrogant attitude.

In the following three features, readers can discover how W.C. Falkner's strategies work to give a positive impression on Southern antebellum culture using the American Revolutionary War's hero and heroine. First is the narrative structure W.C. Falkner uses. The tragic couple's love story during the revolution is told by a centenarian. Oscar Falkland and his lover Olivia Delroy nurture their love in the ideal rural life at Rolla Falkland's Rockland farm in the Hudson River area, but they both die doing their honorable deeds in the American Revolutionary War to protect their land and belongings; their death is lamented "by nearly every soldier in both armies, as well as all the citizens in a radius of a hundred miles from Rockland" (LBC 421). Barnard nostalgically narrates this tragic story of "Revolutionary heroes" (LBC 426) as one of the important episodes during the infancy of the nation.

If it is possible to see Barnard as the substitute for the ex-slave narrator in the Southern story, the narrative structure of *The Little Brick Church* follows the typical pattern Southern authors use in the late nineteenth and early twentieth century. As Wells argues, fiction, nonfiction, poetry, ballads, and folk songs that depict the antebellum plantation life produce the images of slaves or former slaves, who long for the old plantation days, as "cultural clichés."[96] These images' nostalgic tone is effective to paint the past in a positive light. The main narrator of *The Little Brick Church* is a white centenarian instead of an old former slave in order to make the tale of the North nostalgically represented in plantation discourse. The old Northerner's sentimental longing for the good ol' days is positively perceived, even if his story contains embarrassing social structures for contemporary readers.

W.C. Falkner's second tactic is to depict the North's eighteenth-century rural life as if it was another version of the Southern plantation household. The narrator never claims the South is the origin of America, but his rhetoric intentionally suggests the Hudson River area in Colonial America is full of rural elements that are common in Southern plantation life. A hero in the American Revolutionary War is brought up in the landowner's family, whose comfortable life with some domestic slaves has the same structure as the Southern wealthy planter in the antebellum era. This ideological bias should be explained to ESL as well as general students if this text will be properly taught in ESL classes or other undergraduate courses.

The relationship between landowners and their house servants in this novel suggests their social structure in the narrow community is similar to slavery in the antebellum South. When Rolla and Mary Falkland decided to adopt their neighbor's illegitimate girl, Olivia, they found they should also purchase the female slave, Milly, to whom Olivia's mother had trusted her newborn girl on her deathbed. Milly was a slave who was "able to do good work in the field" (LBC 42) and wanted to be with her children, Tilly and Clinky. Rolla decided to "buy the whole family" of Milly (LBC 42). Rolla negotiated with Delroy, who gladly accepted Rolla's offer and was willing to sell Milly and her family to get rid of his shameful granddaughter. Delroy said: "I will give the infernal negroes to you if you will never let me see the bastard any more" (LBC 46). Mr. and Mrs. Falkland expanded their family members to include Olivia and three black slaves.

Barnard suggests Rolla's deal with the slave trader was satisfying not only morally but also economically. A healthy slave family was valuable property for the landowner. The narrator depicts Milly's two children using typical plantation rhetoric to evaluate the healthy slave children:

> Milly's youngest child was a little, fat, healthy boy, whose gold-colored skin denoted the mixture of his blood with that of the Anglo-Saxon race. He was at that time scarcely a year old. Tilly, the daughter, was then in the fourteenth year of her age. She was a bright mulatto, with large expressive eyes and shapely form. There was a sad and thoughtful expression constantly apparent on her features. She had an abundance of long black curly hair. Her smooth yellow skin was of a brighter hue than that of her infant brother. She was rather tall in stature for a girl of fourteen, but her general appearance was attractive, and in some respects beautiful. (LBC 46–47)

The narrator's portrayals of mulatto children are another version of Caroline Lee Hentz's proslavery depiction of the ideal slave child. In *The Planter's Northern Bride* (1854), the generous planter's faithful mulatto boy, Albert, is depicted as a treasured servant with a beautiful appearance: "Albert was a handsome, golden-skinned youth, with shining black hair and eyes, dressed very nearly as genteelly as his master, and who generally attracted more attention on their Northern tour."[97] These aesthetic descriptions of mulatto characters' beauty are safe for the general readers in the nineteenth century because they do not suggest the abuse of slaves in real American history, but insinuate the ideological racial superiority of Anglo-Saxons with their aesthetic judgment.[98] W.C. Falkner uses a typical image of plantation discourse to insist on the value of slaves in the Northern family.

W.C. Falkner uses a third tactic to promote the idyllic condition of plantation and slavery: he designs the slave traders in the North as typical villains and argues the paternal landowner's liberal spirit is inherited by the Southern planters. Barnard describes Donald Delroy and Albert Dunbar, "the commander of a ship that was engaged in the African slave-trade" (LBC 28) and his overseer, as the villains of this novel. Barnard describes how Delroy and Dunbar imported a group of Africans as slaves:

> It has never been my misfortune to behold a more pitiful sight than that presented by the drove of naked and half-starved slaves as they filed past this very door. I noticed that many of their naked backs were covered with bleeding stripes lately made by Dunbar's triple-thonged whip. Their bare feet had been lacerated by the rough stones that covered the road over which they had been driven, and innumerable stains of blood were to be seen on the ground where they had passed. (LBC 66–67)

Blaming the North's slave trading as the origin of the slavery, Barnard defends the Southern slaveholders' position. He argues that "the Southern slaveholder is a saint" compared with the slave traders, who are "heartless demons" (LBC 67). It seems, as Duclos argues, that "speaking through Barnard, [W.C.] Falkner voices the standard 19th century Southern argument excusing slavery in the South by placing the blame on the Northerners,"[99] but actually what Barnard blames are the Northern slave traders. By contrasting the Northern slave traders and the Southern slaveholders, Barnard argues the spirit of liberty, for which early Americans fight, is preserved in the Southern planter's paternalism as well:

> Bear in mind, however, I do not justify slavery in any form. I have shed my blood freely in the cause of liberty. Not for that sort of liberty that protects the strong tyrant while he inflicts his devilish cruelties on the weak and helpless, but the liberty for which I fought was that which I fondly hoped would shield the weak from the oppressions of the strong. The Southern slaveholder does not make slaves out of freemen, but he buys those that the Yankees have converted from freemen into slaves. The greatest crime connected with the institution is committed by those who kidnap them from their homes and reduce them to a state of servitude. . . . As I know of my own knowledge that most of the slave-owners of the South treat their negroes with the tenderest humanity, furnishing them with wholesome food, substantial clothing, and medical treatment when needed, only requiring a moderate quantity of labor. (LBC 67–68)

By placing the villainous roles on the slave traders and Yankees who sell the Africans to wealthy landowners, the South is imagined as a place where the spirit to "shield the weak from the oppressions of the strong" is realized. The author uses Barnard as his mouthpiece to defend Southern slavery and blame the slave traders in the North as the most significant source of depravity. Plantation paternalism for the slaves is approved as the liberal spirit to try and protect the weak from tyrannical dominance.

The ideology that slaves are "the weak" (LBC 67) and should be protected is enforced in this novel's paternalism. The Falkland family in *The Little Brick Church* is organized by the system of patriarchy, in which the white couple creates a complete unity by the wife's obedience to her husband: "Mrs. Falkland was the very sort of a wife that her noble-hearted husband deserved" (LBC 33). And Barnard believed that the landowner couple's "happiness was as near perfection as is ever attained by people on earth" (LBC 33). Belonging to this larger family, Milly's family also swears to obey their masters' will. After Mr. and Mrs. Falkland bought Milly's family from the slave trader Dunbar to their home, Milly and her daughter have a conversation that shows their willing obedience:

> "Oh, mamma!" she cried, "is it true that we belong to Mr. Falkland? Will he not let old Dunbar take us back any more?"
>
> "Yes, honey," replied the mother, smiling through her tears, "Mr. Falkland owns us now, and this is to be our home forever."
>
> "Mamma, do you think heaven is a better home than this?"
>
> "Ah, honey, let us be content with this one until it pleases God to show us into His home. You must be a good girl, and try hard to please your new mistress in everything." (LBC 52)

With their worship for their landowner couple's paternalism, the obedience of Milly and her daughter to Mr. Falkland is naturalized. Mr. Falkland is admired as the person who will protect the weak and helpless slaves, and his position never contradicts their religious spirit:

> It was the custom of the house for each member of the family to rise at the earliest peep of dawn, when the day was opened with Bible-reading and prayer. All the servants, as well as the hired hands, were required to be in attendance at their devotional exercises. Mr. Falkland would read a chapter of the Holy Scriptures, after which all knelt, while a short prayer was offered by him. (LBC 60)

The whole structure of the family members' obedience is illustrated in their everyday Bible reading and prayer.

The white female master's in-home education of the female slave guarantees the beloved family servants' autonomous obedience to this hierarchical order. Mr. and Mrs. Falkland educate Milly's daughter as members of the family. The house slaves are fully educated as capable housekeepers and then voluntarily choose to remain inside the house to keep their position: "Tilly was placed on duty as housemaid, and right well indeed did that faithful girl perform her task. She was quick to learn and prompt in the performance of her labors. She watched her kind mistress [Mrs. Falkland] with anxious eyes, as if she was desirous of anticipating her wishes without waiting to be ordered" (LBC 59). This expanded family structure supported by religion and education gives the practical sense of belonging to Tilly. The sense of belonging brings pride to the slaves, and they are baptized with the aura of the landowner family's class conscious.

The evil slaveholder's revengeful violence against Tilly shows how this family-oriented class conscious is symbolized in the slave's sense of belonging to the Falkland family. When Tilly is fatally forced to return to her slave trader, Dunbar, he cruelly flogs her to humiliate her as his slave again:

> "Now, my proud lady," said he [Dunbar], "I think I'll take a little of them high-flown notions out of you before I'm done with you. Didn't I tell your ladyship that you couldn't put on your grand airs about me? You wouldn't believe me, but I guess you'll do it next time. I knew them Falklands would cram your head full of their aristocratic nonsense, but I'll fetch 'em out of you. You think yourself too grand to associate with a common fellow like me, do you? Very well, I'll teach you better manners. I'll let your ladyship know that I'm as good as anybody. You despise me, do you? Can't love an honest fellow like me! All right, I'll change your notions about that before I'm done with you. But perhaps you have already changed your mind,—what do you say?"
>
> "Despicable wretch," cried the brave girl [Tilly], "I defy you; I spit upon you!" And suiting the action to the word, she spat in his face.
>
> Then he began to apply the lash with the redoubled energy, lacerating the skin at every blow, while the blood trickled from every wound. The poor girl endured the cruel punishment without a groan or a murmur. (LBC 112)

This cruel flogging scene results in Tilly's death. Dunbar's obsessive, violent desire for Tilly suggests Dunbar's sexual perversion and his hatred of the class consciousness vested in Tilly. W.C. Falkner's literary works in the nineteenth century tell readers that an individual's class consciousness is decided by his/her sense of belonging to a family-oriented community. Because of this sense of belonging, the narrow community's social structure curiously undermines the race-based stratification. Even if they are slaves, they can belong to a decent family and can be proud of their membership to a high-class community.

Using the three tactics above, W.C. Falkner describes how a white landowner's patriarchal dominance, which puts various racial and foreign elements under its control, is the foundation of the origin of America. It is certainly not that such a sense of dominance is truly the origin of American mind, but, from today's point of view, this popular romantic author's naturalization of the white male characters' dominance of other socially marginalized members under the liberal spirit enjoys its longevity. In both Fa(u)lkners' works, the class conscious and pride inherent in the decent white family overshadow the people's social distinctions according to race. In W.C. Falkner's fiction, the class consciousness is not decided by one's color of the skin but by one's belonging to a decent white's family-oriented community. This condition is ironically depicted in Pedro's social position at the end of *The Spanish Heroine*. Pedro, whose skin is "a little dark" (SH 104) and his position has been a peon, turns out to be a descendant of a count and becomes a nobleman. And yet he accepts his own subordinate position under Henry's dominance and becomes a member of Henry's community in Kentucky, to establish his honorable position in the United States. In Faulkner's *Intruder in the Dust*, this sense of belonging to a decent white family is recycled in Lucas Beauchamp's pride. Lucas Beauchamp's grandparents were a slave woman and Lucious Quintus Carothers McCaslin, a white planter and slaveholder, and Lucas's pride and sense of belonging to this white decent family is not so much hindered by his racial category. The modern white people's irritation toward Lucas's pride suggests that this sense of social belonging is still effective in the mid-twentieth-century American South. In *Intruder in the Dust*, the strong gravity of the antebellum South's class consciousness surpasses the modern white male's social superiority.

The connection between racial distinction and social status is vague in the nineteenth-century Southern culture. As Martha Hodes reports in *White Women, Black Men: Illicit Sex in the Nineteenth-Century South* (1997), even if "a white woman" engaged with "a black man" in North Carolina, Kentucky, or Georgia, these cases were "the targets of local disdain" but not "the targets of legal action." In antebellum Georgia, census takers in 1860 decided that 8 percent of slaves and 57.2 percent of free African Americans should be classified as "mulatto" rather than "black." One example is Joseph Nunez, whose white neighborhood recognized Joseph's father as a "dark mulatto" and described Joseph as a "mulatto" who had "negro, Indian and white blood in him." Practically, "the Nunezes were slaveowning free people of color."[100] Both fiction and historical facts suggest the family ties and bloodlines, which include extended family members, have a plenty of possible mulatto elements in the Southern families' class conscious.

In the American South, the one-drop rule does not work well to segregate society when compared to the slavery system in the antebellum South. The one-drop rule is quite a vague way to decide people's racial belonging. As W.C. Falkner demonstrates in *The Siege of Monterey*, the white patriarch's decision of his descendant's belonging is the best, most clear solution for racial categories in the family. As Hodes argues,

> The racial caste system that emerged in the post-Reconstruction South was neither slavery nor equality but rather a system of racial subordination based on an ever more rigid color line. Whites ultimately preserved racial hierarchy with the "one-drop rule," whereby anyone with any known African ancestry was

> classified as "Negro." As early as 1876 a black man had observed, "They call everybody a negro that is as black as the ace of spades or as white as snow, if they think he is a negro or know that he has negro blood in his veins." Whereas the institution of slavery had satisfactorily preserved the equation of blackness and slavery, the one-drop rule was the next best solution in a world that lacked legal bondage.[101]

Hodes considers the one-drop rule was not the best solution to decide the racial caste system because this rule did not guarantee the absolute dominance of the white males, who had so long reproduced mulatto descendants under their patriarchal condition that their white descendants may live with their half-brother or half-sister African Americans under the same roof.

In the antebellum South's plantation system, the family member's racial category is supported by the absolute power of the patriarch. The patriarch decides who can be inside his own house as a family member and who belongs to the slave category. This system often allows people who have African American heritage to enter as family members as long as they obey the absolute power of the patriarch. As Thomas Sutpen's attitude in *Absalom, Absalom!* shows, there is no need for evidence of Bon's African blood. Sutpen just demands to control all the elements by his own will in his narrow, family-oriented community, and he finds that Bon, Bon's lineage, and Bon's home land—Haiti—are not proper elements for his family plan. Sutpen finds he cannot completely control Bon's mother and her relatives. He leaves them. Instead, he finds Yoknapatawpha, which symbolically represents the American South, allows him to exercise this complete control of his family members and slaves. In other words, once he loses the historical condition that allows his absolute power to decide someone's race, he lost his power to create his reality. What Sutpen nostalgically sticks to after the Civil War is the antebellum Yoknapatawpha's historical, geographical condition that allows him to retain his power to create his reality and racial category inside of his narrow community. Sutpen's episode suggests that the traditional social structure of the American South has been ideologically enforced by the white male patriarch's imagined power to decide every member's social status arbitrarily.

In the latter part of *The Little Brick Church*, W.C. Falkner describes Oscar's liberal patriarchy as more attractive for Olivia than England's feudal system. After she has grown into a beautiful young woman, Olivia visits England. Olivia meets her real father, Kate Delroy, who is the wealthiest duke in England. Delroy wants to use Olivia's beauty and make her marry Nordane, an affluent English nobleman. But Olivia rejects this offer and returns to America. Thomas Garland, a friend of Oscar, approves of Olivia's decision:

> The daughter of the wealthiest duke in England, the most beautiful woman in Europe, who could have married a prince if she had wished, steps down from her high station, gives up her vast fortune, discards her princely suitors, leaves the most charming palace on the globe, faces the dangers of land and sea to give her hand and heart to an obscure American youth. (LBC 355)

Garland approves of Olivia's love because of her will to reject the noble position that is promised in England. Olivia decides to return to America and marry Oscar because that will give her a stronger sense of belonging than in England. This sense of belonging is more attractive for her when it is compared to the cold-blooded calculation of her aristocratic father who uses his daughter as a bargaining tool. Olivia expects Oscar's patriarchal passion will protect her and fight against the degraded old world.

Olivia is intensely attracted to the pure passion of Oscar who will maniacally construct his narrow community's order and even violently demolish his foes. This amalgamation of love and violence is a quite intriguing element of W.C. Falkner's ideal hero. Oscar's character is depicted as an appropriate one to be an ideal soldier in the budding era of the American Revolutionary War and to be the faithful lover; though, as a surgeon supposes, Oscar's passionate character is sometimes suicidally vicious and uncontrollable:

> He [Oscar] might fight a hundred bloody battles . . . because he knows no such a feeling as fear. But the danger lies in another direction. His passions are strong, violent, and intense. If he were to fall in love, the sentiment would absorb all other feelings, and his whole soul would become involved in the violent passion. If the object of his affections were to die or prove false to him it would endanger his life. If he were to hate an individual it would be such a dislike as to make it dangerous for the party to cross him in any manner. (LBC 359)

Even though his violent character has dangerous aspects, Oscar is approved as a hero who demands to protect his own lover and his belongings against Britain's feudal force with his monomaniac passion. This imperial sense of protection of his belongings against the monarchical force of the other country is depicted as the heroic, liberal spirit in W.C. Falkner's world.

Near the end of this novel, Oscar and Olivia are happily reunited in America, but they die in battle before they marry. Oscar is caught by English soldiers on the battlefield. The English soldiers try to execute him because of Oscar's murder of the English commander. Olivia identifies herself with Oscar and declares, "I mean to share my Oscar's fate, be it what it may" (LBC 405); she struggles to save Oscar, but after she desperately obtains a pardon from English headquarters she dies. Though he gains pardon and release, Oscar learns of his lover's death and dies of shock. Oscar and Olivia are buried in a single casket in "the little brick church" near Oscar's home (LBC 419). Their bodies are settled in "a beautiful double casket" with marriage clothes as an eternal couple that is mourned by soldiers from "each other's side" (LBC 419). Later, the narrator describes how Oscar and Olivia were treated as the foundation of the liberal spirit during the American Revolutionary War. Just below the lovers' grave, "stood a tall marble pedestal, on which appeared a well-executed statue of the Goddess of Liberty pointing her finger directly at the spot where Oscar and Olivia lay" (LBC 426). In W.C. Falkner's novel, the narrator believes the development of unity between Oscar and Olivia realizes the liberal relationship between the planter's son and the daughter of an English aristocrat. However, today's readers can see that even if the couple's bodies are apparently eternalized to suggest they are part of the foundation of a new American society, this liberty includes the white male's desire for the complete dominance of his narrow world's members.

In this romantic story, the author's romanticized egoism is expressed in the imperial impulse of the narrator, the hero's father, and the hero. In Barnard's narrative, Oscar's parents establish their ideal life with their paternal attitude toward slaves, and Oscar becomes his lover's complete protector with his violent passion. W.C. Falkner's monologism unifies these elements by presupposing that the American liberal spirit is the will to protect the weak and helpless from the enemy. Using this romantic discourse, as Duclos suggests, it frequently seems that W.C. Falkner "spent much of his life under the assumption that everything was made for him."[102] Duclos, who demands to see W.C. Falkner benevolently, insists that "we must also acknowledge that he [W.C. Falkner] made things, or contributed to things, to his fellow man,"[103] but W.C. Falkner's fiction shows that the author's sympathy is literally directed to his "fellow" men—the Anglo-Saxon friends and their belongings. In his arrogant imagination, W.C. Falkner builds up his imagined liberal spirit against the reality of the past and present to affirm his own position and imperial attitude, which he hopes will continue eternally. W.C. Falkner's great-grandson, Faulkner, observes that this ideological idea of American liberty is still circulated in his contemporary age, and he criticizes this condition when he depicts Gavin's paternal attitude toward "Sambo" and women in *Intruder in the Dust*.

Gavin's Argument to Reproduce W.C. Falkner's America in *Intruder in the Dust*

In *Intruder in the Dust*, Gavin Stevens's voice retains W.C. Falkner's idea on American liberty that demands to preserve the white male's dominance in a narrow community. Gavin is an attorney from a decent family in Jefferson.

He is a conscientious person who wants to save the old African American, Lucas Beauchamp, from lynching, but once he confronts the possibility of Lucas's social uprising he reveals the Southern elite's imperial personality and desire to control the socially marginalized people, which include African Americans, women, and young people. In the following two monologues, Gavin rhetorically harangues to reestablish the white male's dominance over African Americans and women and tries to persuade Chick with his logic.

In his first argument, Gavin insists that the Southern white male elite is an ideal protector of the African American. Labeling Lucas Beauchamp "Sambo," he insists on the traditional Southerner's paternal attitude to defend his belongings from the outside world and the African American's ability to endure his inferior position:

> I'm defending Lucas Beauchamp. I'm defending Sambo from the North and East and West—the outlanders who will fling him decades back not merely into injustice but into grief and agony and violence too by forcing on us laws based on the idea that man's injustice to man can be abolished overnight by police. Sambo will suffer it of course; there are not enough of him yet to do anything else. And he will endure it, absorb it and survive because he is Sambo and has that capacity; he will even beat us there because he has the capacity to endure and survive but he will be thrown back decades and what he survives to may not be worth having because by that time divided we may have lost America. (ID 199)

Gavin's monologic voice reveals that what he really wants to protect is the white male elite's dominant voice that can control their narrow community. For Gavin, America is synonymous with the self-sufficient, white male elite's dominance. Therefore, if the African American gains a more powerful position that overwhelms the white male adults, his "America" will be "lost." Gavin's curious logic suggests that his desire to dominate and protect the African American is the basic element of the American liberal, an idea that is still part of the modern American South.

Next, he uses the image of an automobile to affirm men's desire to own and control women. In his mind, the white male hero's romantic attitude toward his heroine is connected to modern men's desire for total control of the objects. Gavin's argument reveals his obsession with power:

> The American really loves nothing but his automobile: not his wife his child nor his country nor even his bank-account first (in fact he doesn't really love that bank-account nearly as much as foreigners like to think because he will spend almost any or all of it for almost anything provided it is valueless enough) but his motorcar. Because the automobile has become our national sex symbol. We cannot really enjoy anything unless we can go up an alley for it. Yet our whole

> background and raising and training forbids the subrosa and surreptitious. So we have to divorce our wife today in order to remove from our mistress the odium of mistress in order to divorce our wife tomorrow in order to remove from our mistress and so on. As a result of which the American woman has become cold and undersexed; she has projected her libido onto the automobile not only because its glitter and gadgets and mobility pander to her vanity and incapacity (because of the dress decreed upon her by the national retailers association) to walk but because it will not maul her and tousle her, get her all sweaty and disarranged. So in order to capture and master anything at all of her anymore the American man has got to make that car his own. (ID 233)

Gavin's words rhetorically demonstrate that what the American (he generalizes the Southern male's traditional mindset as an American one) loves is his complete dominance of objects, and he does not care for his wife, child, country, and economy because these cannot be completely controlled by his own will. Things sexually stimulate and jolly him along only when they go up his alley. He regretfully finds that his female partner also wants power and superiority as well. Divorce and materialism instead of loving unity and stoicism become modern society's main streams. Modern women's hearts and souls cannot be dedicated to their male partners; Gavin feels that "the American woman has become cold and undersexed." Gavin's manly mindset that gravitates to the American South's antiquated romanticism finds the automobile's close relationship with the American woman is symbolically connected to the tragic power relationships between men and women in modern America. Yet his outdated idealism prevents him from a new solution. In his argument, a man's tragic desire to control a woman is converted to his monomaniac desire to control his own automobile.

Gavin's attitude is similar to the hero's patriarchal attitude in W.C. Falkner's *The Little Brick Church*. His monologic voice confesses that the liberal, modern Southern elite still retains antique romanticism and has a strong desire to control others. Gavin, who wants to keep his own self-sufficient, white male position, insists on his preference for the complete protection and control of African Americans and women using the metaphor of "Sambo" and an automobile.

In his dialogue with Gavin, Chick personally struggles to emancipate himself from the ideological gravity of the modern American South's imperial impulse. He has to reconfigure the image of his own self, reflecting on his own personal connections with the minor members of the society: Lucas, young African American Aleck Sander, and the spinster Miss Habersham. These socially marginalized people's critical views on white male dominance give Chick chances to stand apart from his uncle's monologic masculinity. *Intruder in the Dust* reveals that W.C. Falkner's gravity can be critically observed from the eyes of youths, African Americans, and females. Chick shifts his imagined identity from an independent, self-sufficient person to one who finds his own intersubjective existence among his personal group, "*me and Aleck Sander and Miss Habersham*" (ID 205).

Chick critically sees that the American South is represented by one monologic persona based on the white male's aggressive will to protect their community. For him, the white male community is "a Face, the composite Face of his native kind his native land, his people his blood his own with whom it had been his joy and pride and hope to be found worthy to present one united unbreakable front to the dark abyss the night" (ID 190). Once Chick can imagine that he will take part in this dominant group, he is integrated to be the person who protects this monolithic community. Chick knows that the power of "one united unbreakable front" is formed by such individualistic and fetishistic reliance on the romantic identity and unification, and he still feels that his mind is attracted to such a romantic unity with the monolithic community's illusion. Chick's personality partly shares this stubborn doctrine of the white male Southerners. Chick feels that this image of the community is "his native kind his native land," and "his joy and pride and hope," and that it can be found in "the composite Face" of white males. This image of the typical white male's "Face" inherits W.C. Falkner's romanticism. As Oscar's violence toward the enemy and his obsessive desire to control his woman and his slaves is shown, W.C. Falkner's idealized white male image aggressively defends his own narrow community against his foes. In *Intruder in the Dust*, Chick explores a way to resist this pattern and establish an alternative relationship with multiple community members.

Gavin's arguments about "Sambo" and women can be interpreted as Faulkner's critical mocking of his great-grandfather's authoritative voice. W.C. Falkner's imagination portrays American history as if it could be represented by some hero's violent will to protect his community. Faulkner's white male characters in his early and middle works are attracted to this monolithic vision because of their desire to keep their own self-centered world. Gavin also shares this strong tendency as a white male elite in Yoknapatawpha. However, Faulkner creates Chick in *Intruder in the Dust*, a character who can listen to the voices of others—African Americans and women—and cooperate with them. In *Intruder in the Dust*, Chick shares food with Lucas and hears Lucas's demand. Chick becomes a person who can carry the penetrated word in his own discourse. As Bakhtin says, a character's "*penetrated word* . . . is, a word capable of actively and confidently interfering in the interior dialogue of the other person, helping that person to find his own voice" (PDP 242). Chick gives a voice to Lucas's intentionally silenced words, and gains the ability to democratically combine his own and the other's voices. In this process, Chick finds an intersubjective viewpoint with Lucas to resist Gavin's authoritative logic. Faulkner shows the young protagonist's ability to cooperate with others to improve their condition. This development from W.C. Falkner to Faulkner is a gradual progress from monologism to dialogism, and from authoritative discourse to internally persuasive discourse.

In *Intruder in the Dust*, the socially marginalized characters appear to destabilize the local community's future by their social combinations, condemning the sadistic corruption of the white males' self-sufficient community. In spite of its various heterogeneous elements, W.C. Falkner's romantic vision concludes that only the white males' monologic discourse can give order to the narrow community. What Faulkner criticizes and fights against is this monologic discourse. Faulkner's characters, families, and communities show a historical development of the literature of American South—the disintegration of patriarchal monologism. Faulkner's progressive demand for an alternative formation of society begins with *Intruder in the Dust*, which depicts socially stratified individuals' dialogical relationships in their struggle to realize their personal democracy. For ESL students, W.C. Falkner's texts can be used to demystify Faulkner's works. By comparing these two Fa(u)lkners' works, ESL students can understand what kind of idealistic pattern the white male characters share. The imperial power of white male adults is sustained in the people's imagined worldview and practically conducted in racial segregation and social stratification in the American South. If these social structures are studied correctly, ESL students can understand what kind of gravity dominates this community.

CONCLUSION

W.C. Faulkner might be a second-rate writer, but his romanticized heroes are important for understanding the historical origin of the modern characters' obsessions in Faulkner's works. Comparing both Fa(u)lkners' works helps readers comprehend Faulkner's struggle to reconfigure Southern discourse from monologism to dialogism. In W.C. Falkner's works, the white male Southerner dominates their community with absolute power. W.C. Falkner's works are full of struggles with foes, confusions of social status, and threats of racial mixing, but his white male protagonists finally create order again under their domination. W.C. Falkner ideologically settles this patriarchal attitude of white males as the origin of Southern plantation culture. Because his great-grandfather's works show a typical pattern of the Southern white male's imagined worldview, Faulkner—a modern author in the twentieth century—re-accentuates W.C. Falkner's motifs to critically show that the patriarchal discourse of the Southern white males still has power in the modern American South.

W.C. Falkner's monologic worldview is still worth studying because it tells us precisely where Faulkner's mythic world comes from. Faulkner's early and middle works focus on how the white male character's psychological agony tragically fails to resist W.C. Falkner's idea of romantic unity. Faulkner's early protagonists imagine the ultimate unity between the self and death (*Soldier's Pay, Flags in the Dust*), the self and ideal design (*The Sound and the Fury, Absalom, Absalom!*), and the self and nature (*Go Down, Moses*), disregarding or excluding their social combinations with others. In this reading we can view Faulkner, not as the great representative or "official language"[104] of the American South, but as a part of the process to reconfigure the literary language of the region. In this process of reconfiguration, Faulkner's works can be considered a part of the global world's cultural process from romantic monologism to heterogeneous dialogism.

Monologism seems to be one of the most effective ways to declare a subject's future direction. When we consider our future, we tend to bind our minds with our own words, and we believe our iron wills can help us attain our purposes. But Thomas Sutpen's life in *Absalom, Absalom!* suggests excessive monologism overlooks the broad diversity of multiple connections based on democratic relationships. In Sutpen's monologic design, a person's successful life causes the victimization of other social members and prevents people's fair evaluation of their social combinations. As Shreve's attitude shows in *Absalom, Absalom!*, international society encourages the prosperity of people's dialogical relationships; however, some characters who are obsessed by the romantic hero's successful story of the past tend to avoid (or, like Quentin, only say "no" to) dialogue and fall back into monologism. Faulkner's early and middle works show the white elite protagonists' obsession with the power of monologism, which creates and certifies their self-content ambition, even if it is an illusion.

Comparing W.C. Falkner's works, *Siege of Monterey*, *The Spanish Heroine*, *The White Rose of Memphis*, and *The Little Brick Church* with Faulkner's works, I argue that W.C. Falkner's heroic image and imagined worlds are important elements in Faulkner's early and middle works. Examinations of both Fa(u)lkners' works prove that Faulkner's early and middle works criticize W.C. Falkner's romantic motifs; however, Faulkner's early and middle works do not offer an alternative worldview to resist W.C. Falkner's self-centered perspective. Since the main characters of Faulkner's early and middle works cannot focus on their intersubjective worldviews with others, they are tragically obsessed with W.C. Falkner's idea of romantic unity under the dominance of the white male heroes. When the main characters in Faulkner's early and middle works demand that their eternal worlds be governed by their own wills, they are caught in the centripetal force of W.C. Falkner's romantic imagination. *Intruder in the Dust* is a breakthrough in Faulkner's worldview. The protagonist in *Intruder in the Dust*, Chick, successfully resists this monologic gravity. In the next chapter, I expand my argument on Faulkner's *Intruder in the Dust* to analyze how Faulkner depicts social combinations of socially marginalized members in the Southern community to reconfigure the modern American South as a place where individuals can form egalitarian relationships and keep their community's changeability and how the American literature teacher can explain this analysis to ESL students.

NOTES

1. William C. Falkner, *The Siege of Monterey: A Poem* (Cincinnati: By the Author, 1851), 3.
2. William Faulkner, *The Marble Faun*, in *The Marble Faun* and *A Green Bough* (New York: Random House, 1960), 12.
3. Michel Foucault, "What Is an Author?" in *The Foucault Reader*, ed. Paul Rabinow (New York: Pantheon, 1984), 107, 119.
4. Ibid., 118.
5. Richard Godden, *Fictions of Labor: William Faulkner and the South's Long Revolution* (Cambridge: Cambridge University Press, 1997), 4.
6. This sentence is hinted by de Man's comment on Derrida's contribution to Rousseau studies. See Roy Boyne, *Foucault and Derrida: The Other Side of Reason* (London: Routledge, 2001), 120n6.
7. Jacques Derrida, *Of Grammatology*, trans. Gayatri Chakravorty Spivak (Baltimore: Johns Hopkins University Press, 2016), 65, 155–56.
8. Jere Hoar reports that, since book reading has not generally been characteristic of adults in the United States, he researched the practices and preferences that could be expected from an aging population. In an attempt to answer this question, the reading habits and preferences of persons sixty years of age and over in a rural Mississippi town were investigated. The 200 persons interviewed were randomly selected. Hoar finds that "the only books mentioned by more than one respondent as having been the last ones read were: *This Hallowed Ground*, *The Nun's Story*, *The Holy Bible*, *The Bridge at Andau*, *The Last Angry Man*, *The White Rose of Memphis*, *The River Road*, *The Scapegoat*, *Imperial Woman*, *Readers Digest Condensed Book*, and *Intruder in the Dust*." Apparently, some ordinary readers in Mississippi could easily remember W.C. Falkner's work even in the middle of the twentieth century. See Jere Hoar, "Book Reading in the Senior Years: The Habits and Preferences of 200 Mississippians," *Journal of Educational Sociology* 34, no. 3 (1960): 140.
9. Ibid., 138.
10. Robert Cantwell, "The Faulkners: Recollections of a Gifted Family," in *William Faulkner: Three Decades of Criticism*, ed. Frederick J. Hoffman and Olga W. Vickery (New York: Harcourt, 1963), 55.

11. Frederick J. Hoffman, "William Faulkner: An Introduction," in *William Faulkner: Three Decades*, 27.

12. As Richard H. King argues, the Southern Renaissance is a cultural revival, in the late 1920s–1940s, of the writers and intellectuals of the South who attempted to come to terms with the "ambivalence and ambiguity" in their region. See Richard H. King, *A Southern Renaissance: The Cultural Awakening of the American South, 1930–1955* (Oxford: Oxford University Press, 1980), 7.

13. Lawrence H. Schwartz argues the critical opinion of Faulkner's works was established when *The Portable Faulkner* was published in 1946, and the commercial revival of Faulkner comes in response to the successful publication of *Intruder in the Dust* in 1948. See Lawrence H. Schwartz, *Creating Faulkner's Reputation: The Politics of Modern Literary Criticism* (Knoxville: University of Tennessee Press, 1990), 26.

14. Mark Royden Winchell, *Cleanth Brooks and the Rise of Modern Criticism* (Charlottesville: University Press of Virginia, 1996), 310.

15. Cleanth Brooks, *William Faulkner: The Yoknapatawpha Country* (Baton Rouge: Louisiana State University Press, 1990), 368.

16. Malcolm Cowley, introduction to *The Portable Faulkner*, ed. Malcolm Cowley (New York: Viking Press, 1974), xi.

17. Joel Williamson, *William Faulkner and Southern History* (New York: Oxford University Press, 1993), 27.

18. In his two-volume biography of Faulkner, Blotner cites Duclos's "thoroughgoing work" as "the best single source of information" on W.C. Falkner. See Joseph Blotner, *Faulkner: A Biography* (New York: Random House, 1974), 1: Notes 7.

19. Duclos, *Son of Sorrow*, 9, 307.

20. Todorov, *Mikhail Bakhtin*, 107.

21. Van Wyck Brooks, *The Times of Melville and Whitman* (Scranton: E. P. Dutton, 1947), 334.

22. For the details of W.C. Falkner's statue and its troublesome episode, see Williamson, *William Faulkner and Southern History*, 62–64.

23. Thomas Felix Hickerson, *The Falkner Feuds* (Chapel Hill, NC: Colonial Press), iv.

24. Library of Congress, http://www.loc.gov/; Internet Archive, https://archive.org/.

25. Google Books, http://books.google.com.au/.

26. The New York Public Library, https://www.nypl.org/; HathiTrust digital library, http://babel.hathitrust.org/.

27. Ibid.

28. Duclos, *Son of Sorrow*, 151.

29. No official organization offers the full text of *The Little Brick Church*, but personally, in order to share the text with my students, I published the digitized text of *The Little Brick Church* on the web; see https://ltsproject.wordpress.com/.

30. John B. Padgett, *William Faulkner on the Web*, University of Mississippi, 1995, http://www.mcsr.olemiss.edu/%7Eegjbp/faulkner/faulkner.html.

31. Stephen Railton, *Digital Yoknapatawpha*, The National Endowment for the Humanities; The Department of English, UVA; The Office of the Dean, College of Arts & Sciences, UVA, http://faulkner.iath.virginia.edu/.

32. William Faulkner. *The Sound and the Fury: A Hypertext Edition*, ed. R. P. Stoicheff et al., Updated February 2011, University of Saskatchewan, http://drc.usask.ca/projects/faulkner/; Dirk Van Hulle, "Hypertext and *Avant-texte* in Twentieth-Century and Contemporary Literature," in *A Companion to Digital Literary Studies*, ed. Ray Siemens and Susan Schreibman (Malden, MA: Blackwell, 2007), 140.

33. Stephen Railton, *Faulkner at Virginia*, University of Virginia, http://faulkner.lib.virginia.edu/.

34. V. W. Brooks, *The Times of Melville and Whitman*, 331.

35. Cantwell, "The Faulkners," 55.

36. Bondurant claims he gained "the valuable information" from W.C. Falkner's son J.W.T. Falkner, W.C. Falkner's daughter Mrs. N.G. Carter, children of W.C. Falkner, and his lifelong friend, Mrs. M.H. Crockett. See Alexander L. Bondurant, "William C. Falkner, Novelist," in vol. 3 of *Publications of the Mississippi Historical Society*, ed. Franklin L. Riley (Oxford, MS: Mississippi Historical Society, 1900), 113.

37. Ibid., 121.

38. Duclos, *Son of Sorrow*, 9.

39. Bondurant, "William C. Falkner, Novelist," 115–17, 120.

40. Ibid., 120–22.

41. Ibid., 122, 125.

42. Joseph Blotner, *Faulkner: A Biography*, one-volume ed. (New York: Vintage, 1991), 22.

43. Robert Cantwell, "Introduction," in *The White Rose of Memphis*, by William C. Falkner (New York: Coley Taylor, 1953), v, xiv, xv.

44. Ibid., xxiv–xxv.

45. Ibid., xxv.

46. Duclos, *Son of Sorrow*, 249.

47. Hickerson, *The Falkner Feuds*, iv, 8; Williamson, *William Faulkner*, 60.

48. Williamson, *William Faulkner*, 62–64; Hickerson, *The Falkner Feuds*, 28–29.

49. For further information on W.C. Falkner and the US-Mexican War, see Duclos, *Son of Sorrow*, 39; Hickerson, *The Faulkner Feuds*, 11; Williamson, *William Faulkner*, 18. For further information on W.C. Falkner and the Civil War, see Duclos, *Son of Sorrow*, 101; Hickerson, *The Faulkner Feuds*, 24; Williamson, *William Faulkner*, 41.

50. *The Memphis Appeal* (Memphis, Tennessee) July 26, 1861, quoted in Duclos, *Son of Sorrow*, 101.

51. Duclos, *Son of Sorrow*, 121; Hickerson, *The Faulkner Feuds*, 25; Williamson, *William Faulkner*, 42–43.

52. Williamson, *William Faulkner*, 43.

53. Duclos, *Son of Sorrow*, 145; Hickerson, *The Faulkner Feuds*, 27–28; Williamson, *William Faulkner*, 46.

54. Bondurant, "William C. Falkner, Novelist," 116.

55. Duclos, *Son of Sorrow*, 53.

56. Ibid., 366; Williamson, *William Faulkner*, 19.

57. Williamson, *William Faulkner*, 19.

58. Duclos, *Son of Sorrow*, 53.

59. Williamson, *William Faulkner*, 37.

60. Ibid., 49.

61. Bondurant, "William C. Falkner, Novelist," 121. Duclos argues that "the great legend of [W.C.] Falkner as a railroad builder" has been "perpetuated by Bondurant, Winston, Cantwell, Coughlan, the present generations of the Falkner family, and even by two railroad historians: Stewart Holbrook and J. H. Lemly." See Duclos, *Son of Sorrow*, 153.

62. Duclos, *Son of Sorrow*, 153; Hickerson, *The Falkner Feud*, 30.

63. See Andrew Brown, "The Railroad," quoted in Hickerson, *The Falkner Feud*, 30.

64. Duclos, *Son of Sorrow*, 154.

65. Ibid., 152–55; Hickerson, *The Falkner Feuds*, 30–31.

66. Duclos, *Son of Sorrow*, 236.

67. Ibid., 235–39; Hickerson, *The Faulkner Feuds*, 31.

68. Williamson, *William Faulkner*, 25–29; On Williamson's discovery, Doyle comments as follows: "Joel Williamson's study of Faulkner's great-grandfather Colonel William C. Falkner reveals that some mulatto descendants of a woman who once was his slave believed he was their ancestor. Whether this was true, and whether this was something the family even whispered about, is not known. Given the prevalence of mulattoes in Mississippi, any white person (whether or not their family had once owned slaves) might wonder if someone across the color line might be kin." See Don H. Doyle, *Faulkner's County: The Historical Roots of Yoknapatawpha* (Chapel Hill: University of North Carolina Press, 2001), 149.

69. Williamson, *William Faulkner*, 55–56.

70. William Van O'Connor, *The Tangled Fire* (New York: Gardian Press, 1968), 9.

71. Carl Guarneri, *America in the World: United States History in Global Context*, 157.

72. Blotner, *Faulkner*, 1: 345, 1: 361.

73. Duclos, *Son of Sorrow*, 296

74. V. W. Brooks, *The Times of Melville and Whitman*, 332.

75. Stark Young, "Golden Apples," *English Journal* 26, no. 9 (1937): 690.

76. Ibid., 691.

77. Joep Leerssen, "Literary Historicism: Romanticism, Philologists, and the Presence of the Past," *Modern Language Quarterly* 65, no. 2 (2004): 222.

78. Walter Scott, *Ivanhoe* (New York: American Book Company, 1904), 102, 136–37.

79. Duclos, *Son of Sorrow*, 84.

80. Falkner, *The Siege of Monterey*, vii.

81. Blotner, *Faulkner*, 1: 16.

82. Ibid., 1: 17.

83. Ibid., 1: 18, 1: 15.

84. Hickerson, *The Falkner Feuds*, 11.

85. Marta Puxan, "Narrative Strategies on the Color Line: The Unreliable Narrator Shreve and Racial Ambiguity in Faulkner's *Absalom, Absalom!*" *Mississippi Quarterly* 60, no. 3 (2007): 531.

86. Noel Polk, *Children of the Dark House: Text and Context in Faulkner* (Jackson: University Press of Mississippi, 1996), 138–39.

87. Puxan, "Narrative Strategies on the Color Line," 551.

88. Blotner, *Faulkner*, 1:18.

89. Robert Cantwell, "Introduction," in *The White Rose of Memphis*, by William C. Falkner (New York: Coley Taylor, 1953), viii.

90. Duclos, *Son of Sorrow*, 1.

91. Jeremy Wells, *Romances of the White Man's Burden: Race, Empire, and the Plantation in American Literature, 1880-1936* (Nashville: Vanderbilt University Press, 2011), 180, 178, 183.

92. Wells, *Romances,* 7.

93. Ibid.

94. Ibid., 184.

95. Thomas Dixon Jr., *The Leopard's Spots: A Romance of the White Man's Burden, 1865-1900.* (New York: Doubleday, Page and Company, 1902), 159, https://archive.org/details/leopardsspotsrom00dixo.

96. Wells, *Romances,* 3.

97. Caroline Lee Hentz, *The Planter's Northern Bride* (Philadelphia: T. B. Peterson and Brothers, 1854), 14.

98. The aesthetic judgment of the characters' racial appearance that supported the Anglo-Saxon's racial superiority made these novels acceptable to both readers in the North and the South during the nineteenth century. As Diane N. Capitani argues in *Truthful Pictures* (2009), a physiognomy study in the nineteenth century argues "the degree of blackness in the skin color . . . seemed to be a determining factor in the African's lack of progress," and in popular romances in the nineteenth century, "North and South, whites had an extremely difficult time finding beauty in the black race." Many Southern popular novels reinforce this idea, and "even in Mrs. Stowe's work [*Uncle Tom's Cabin*] it is evident that those slaves that are the most attractive are usually mulattoes." These ideological discourses presuppose that the beauty is caused by the Anglo-Saxon blood. See Diane N. Capitani, *Truthful Pictures: Slavery Ordained by God in the Domestic Sentimental Novel of the Nineteenth-Century South* (Lanham: Lexington Books, 2009), 125.

99. Duclos, *Son of Sorrow*, 201.

100. Martha Hodes, *White Women, Black Men: Illicit Sex in the Nineteenth-Century South* (New Haven: Yale University Press, 1997), 52–53, 98, 101, 104.

101. Ibid., 199.

102. Duclos, *Son of Sorrow*, 215.

103. Ibid.

104. Kreyling, *Inventing Southern Literature*, 127.

Chapter 2

Social Combinations beyond Borders of Race, Gender, and Class in William Faulkner's *Intruder in the Dust*

> When Faulkner met some of the citizens of Nagano, Japan, in 1955, they asked him which of his books they should read first. Faulkner replied, "I would suggest—the title is *Intruder in the Dust.* I suggest that because that deals with the problem which is important not only in my country, but, I think important to all people."
>
> —Patrick H. Samway, S. J.,
> *Faulkner's "Intruder in the Dust"*[1]

Teaching William Faulkner's text in an ESL class is a task that reexamines the American South's modern literature in today's globalized contexts and digital environments. It means, practically, that a teacher talks about Faulkner's texts while sitting in front of so-called nonnative students who have smartphones, tablets, or computers within their reach. Many students use their online dictionaries to check the meaning of English words and phrases. They tap in some keywords and then search engines give them cultural images immediately. Various social networking interfaces—including Twitter, Facebook, Instagram, and YouTube—give opportunities for all class members to be connected to their friends in a digital space in or outside of class. Today's multimedia classroom is generally equipped with technologies—a DVD and Blu-ray player, a computer, a projector and a screen, the university intranet's Moodle system, and Google Drive—that can be used to distribute, gather, or share text, files, images, and videos.

All of these conditions of undergraduate education mean that teachers' and students' dialogical relationships are more objectively displayed in digital media. Teachers know that their writings on the board, or even their utterances in class, are snapshotted and texted by students' digital tools and uploaded to their online social networks. Students' writing and their everyday events are also documented and captured in the various forms—on Moodle discussion boards, short texts on SNS (Social Networking Service), timelines on Facebook, and other texted or visual modes. People's digitized identities are displayed on the net for other audiences. As Barton and Lee argue in *Language Online* (2013), "identity online is not just about *who we are*, but also *who we want to be to others*, and *how others see us*."[2] Students are haunted, or even cursed, by the eyes of others in the digital space. Whether the ESL class is conducted in the Eastern or the Western part of the world, when a teacher uses Faulkner's work as a class text, it means the teacher will discuss how Faulkner's local characters symbolically represent the conditions of students who are very conscious of their own identity management in digital space.

This chapter explores how a teacher can analyze Faulkner's work for ESL students' comprehension. Putting Faulkner's work in front of ESL students implies that Faulkner's work is a metaphor for today's world and identities. The students who are very conscious about their appearances, utterances, and actions in the digital space share similar experiences with Faulkner's Southern characters who are, as Sylvia Jenkins Cook argues, very conscious of their "self-display" and "self-fashioning."[3] It is beneficial for the American literary teacher to analyze the words and actions of Faulkner's characters in an ESL class because Faulkner's characters are typical examples of the formation of people's mutually interwoven identities in a deeply stratified society. Today, ESL students do not witness the lynched body hanging from a tree in their town, but they do witness the contemporary world's threatening of indiscriminate terrorism on YouTube. Instead of the argument for racial segregation in a local area, they read anonymous hate speech on Twitter. White male dominance in the American South is not a familiar topic for ESL students in Japan, but it can be a metaphor for some groups' monopolization of power and money in their globally networked world. The digital network seems to personally connect people beyond borders of class, race, gender, religion, and nation, but it also allows some companies and political powers to expand their dominance to the rest of the world.[4] This global condition can be compared to the social condition of the American South, whose literature impressively depicts its community's personal connections that have expanded beyond seriously stratified conditions according to race, gender, and class under the white male dominance. Expanding John Egerton's idea of "Southernization,"[5] I suggest that in today's globalized society the world and the American South share the worst aspects of both societies: racial segregations; violent domination of some particular group; social stratifications caused by religion, economy, and politics; killing individuals without legal trials. And yet at the same time, the literature of the American South suggests that the people have ability to have equal relationships temporarily in any situation. The modern authors in the American South have studied these terrible aspects and described some bits of hope. If it is possible to say that today's digital world is Southernized, Faulkner's text can be reconsidered as a contemporary tool by which we can analyze what we are now involved in and how to alternate it. Our imagination—based on today's digital experience—can comprehend Faulkner's literary concept, which indicates that local matters directly connect to global issues. As Kreyling suggests, it is possible for today's readers to be conscious of the literature of the American South with their own "body and mind."[6] Faulkner puts this small society's race problem and the fear of lynching in front of world readers and explores how his mutually interwoven characters reflect contemporary readers' relationships with others. This chapter explores how an ESL teacher can talk about Faulkner's work as a text that is suitable for today's students.

What Faulkner focuses on in this novel is the characters' shared words that are powerful enough to change the character's life. In one of the early scenes of *Intruder in the Dust*, a young white protagonist recognizes his social position's uneasy shift occurs when he finds his words dialogically connected with the other. An old African American, Lucas Beauchamp, who is wrongly jailed for the murder of a poor white man, is afraid of being lynched by the white locals. Lucas hopes that a sixteen-year-old white boy, Charles (Chick) Mallison, will prove his innocence, because he had communicated with Chick several times since they shared food four years before. When Lucas requests to dig up Vinson Gowrie's grave to find that it is not the bullet of Lucas's forty-one Colt that killed Vinson, Chick hears his own voice screech with outraged disbelief: "Me? *Me?*" (ID 66). Chick cannot control his own voice and blurts out: "Me go out there and dig up that grave?" (ID 67). Lucas's reliance on his personal connection is an awful signal to the white boy, for it is the beginning of the disintegration of Chick's peaceful world. Chick believed he was separate from Lucas because "he was free" (ID 27), but when he hears Lucas's words in jail he finds he cannot decide his social actions by himself. It is the young protagonist's turning point from the subjective recognition of his independent self to his objective acknowledgment of the uneasy relationship between Lucas and himself. Once he focuses on the language's intersubjective element that reflects not only his intention but also Lucas's will, Chick learns his action does not come out of his own solitary will, but it emerges from his relationship with the other. What today's readers find in *Intruder in the Dust* is this dialogical space that connects Chick's and Lucas's words to generate their contemporary worldview.

Because his social connection with Lucas controls his self and voice, Chick sees himself from his imagined intersubjective viewpoint. Once a social combination has been established among individuals, the voice of each participant appears as a mutual phenomenon that cannot be completely possessed by one individual. This linguistic condition well epitomizes Bakhtin's idea of the dialogized relationship of individuals. According to Bakhtin, individuals' utterances strongly resonate with the voices of others: They are "a series of complex interrelationships, consonances and dissonances" (DN 282), and each "word in language is half someone else's" (DN 293). Each person can have an ability to have equal relationship with the other for a while to learn much. Chick finds that Lucas gives particular context to their words, and their dialogic relationship suggests an action for Chick to perform. Using language, individuals "are capable of attracting its words and forms into their orbit by means of their own characteristic intentions and accents, and in so doing to a certain extent alienating these words and forms from other tendencies, parties, artistic works and persons" (DN 290). Lucas communicates with Chick to change his own destiny from a common routine of the community.

This chapter applies a term, social combination, to reconsider the value of the characters' intersubjective relationships in Faulkner's *Intruder in the Dust.* Social combination refers to two or three individuals' interpersonal connections and efforts to have a personally democratic relationship. Using this term, *Intruder in the Dust* is explored as a work that reconfigures the modern American South as a place in which the society's deeply stratified condition is always intruded upon and undermined by dialoguers' interwoven relationships. The characters create their own personal contexts and vocabularies to make sense of their relationships in a deeply stratified society or in "the social heteroglossia" (DN 370). Chick's social combination with Lucas reconfigures the Southern community, not as a place for people's mutual independence and violence, but as a place for the formation of dialogic space among socially stratified members. Marginalized characters' intersubjective connections in the American South can give an alternative model of social formation under the pressure of a particular group's imperial discourse to international readers who feel social stratifications still exist in people's lives.

Faulkner was proud of his achievement when he completed *Intruder in the Dust.* His evaluation of his work continued positively enough to recommend it to international readers. As Patrick Samway informs, shortly after he completed it, Faulkner considered *Intruder in the Dust* as his best:

> After its publication on September 27, 1948, Faulkner thought highly of the book [*Intruder in the Dust*], at least for a while. When Mrs. Mary Betsy Waddle, a long-time friend of the Faulkners, visited Rowan Oak, the Faulkner home in Oxford, Mississippi, in the late 1940's, Faulkner presented her with a copy of *Intruder* and told her, "I consider this book . . . the best I have ever written."[7]

Samway states that Faulkner's high evaluation continued "at least for a while," but it seems that his confidence on this work was sustained for several years. When Faulkner visited Japan to take part in the Nagano Seminar in

August 1955—under the auspices of the Exchange of Persons Branch of the US State Department—one of Nagano's citizens asked him which one of his works should be read first. After complimenting the Nagano citizen's good English, Faulkner recommended *Intruder in the Dust*: "I suggest that [novel]," he continued, "because that deals with the problem which is important not only in my country, but, I think important to all people."[8] Seven years after the completion of this novel, Faulkner still felt what he had achieved in *Intruder in the Dust* could be supremely recommended to international readers.

Faulkner's early and middle works, which were written before *Intruder in the Dust*, focused on the white male characters' internal agonies in order to criticize their logic of exclusions of the others. From *Soldier's Pay* to *Go Down, Moses*, Faulkner portrayed the white characters' personal obsessions and scarcely showed their possibility to have dialogic relationships with others to create common ground. Specifically, some socially marginalized people's struggles to have relationships with the white elite family members came to tragic ends. Near the end of *The Sound and the Fury*, the black servant Dilsey stroked the idiot Benjy's head and then claimed, "You's de Lawd's chile, anyway. En I be His'n too, fo long, praise Jesus," but Benjy disregarded Dilsey's warm embrace, concentrating on his own obsession and memory, and "wailed steadily" (SF 1121). Dilsey's will to have a relationship with Benjy was disregarded by Benjy's concentration on his world that wanted to keep each item "in its ordered place" (SF 1124). In *Absalom, Absalom!*, a poor white farmer, Wash Jones, believed he gained the possibility to have an rightful relationship with a white plantation owner, Sutpen, when his granddaughter had Sutpen's child, but he was insulted by Sutpen because his granddaughter had a girl instead of a boy. Wash killed Sutpen with a desperate anger at Sutpen's arrogant contempt—Sutpen's disdain destroyed Wash's hope to take part in Sutpen's family-oriented community. In *Go Down, Moses*, old Ike found that an African American girl, who was a descendant of his own ancestor and a slave, loved and pursued Ike's white relative, young Edmond. Ike was shocked by this affair and told her to leave their community and go north to marry "a man in your own race" (GDM 346). The girl accused Ike of misunderstanding her unstoppable love for Edmond: "Old man, . . . have you lived so long and forgotten so much that you dont remember anything you ever knew or felt or even heard about love?" (GDM 346). In the end, Ike's demand for racial purity and the demolition of the interracial relationship was realized by his young relative's cruel purge of his lover. Young Edmond shot her to make an end of their love, pretending that he mistook her for "a deer" (GDM 348). In these works, some socially marginalized characters demanded personal relationships with white male protagonists, but their wishes were neglected or rejected by those protagonists, who demanded to keep their self-content positions or their personal privileges.

In contrast to his early and middle works, in *Intruder in the Dust*, socially marginalized characters' personal combinations attain a positive result in the local community. It took several years for the author to achieve this expression. After publishing *Go Down, Moses*, Faulkner did not publish a novel for six years. What he did was cooperate with Malcolm Cowley to publish *The Portable Faulkner* (1946), intermittently work on his "big book," *A Fable* (1954), and write a few short stories.[9] During these silent years, though, Faulkner's idea for his next novel expanded. His letters in 1940 suggest that the story was originally planned as "a blood-and-thunder mystery novel which should sell" (SL 122), and turned out to be *Intruder in the Dust*. Faulkner evaluated this novel as "a pretty good study of a 16 year old boy who overnight became a man" (SL 266). The author's will "to propel white shame and guilt about racial oppression toward constructive purposes"[10] was reflected in the development of the young protagonist's social combination with the marginalized characters. In *Intruder in the Dust*, Faulkner wrote a more hopeful plot in order to see people's intersubjective recognitions in the American South as a motive for recomposing the community's structure. Today's ESL students, who create multiple connections on a digital space, can see Chick's dialogic relationships with others as a more sympathetic social model for them.

Intruder in the Dust has not been considered one of Faulkner's masterpieces in the history of Faulkner criticism. Many critics consider it a regional novel that narrowly focuses on the white protagonists' agony over the inherent racial problem of the American South, and some early reviewers read it as a manifestation of the Southern white elite's conscience.[11] Two main characters of this novel—Chick, a sixteen-year-old white boy who saves an elderly African American from a lynch mob, and Gavin Stevens, his instructive uncle and an elite lawyer—are considered representatives of the Southern whites' noble spirit that declares a passionate will to aid the African American's fraught condition.

The critics in the middle of the twentieth century portray *Intruder in the Dust* as one of Faulkner's flawed novels, considering Gavin the awkward protagonist of this novel. Disdaining this white elite lawyer who demands to preserve the white male dominated community, three main critics in the 1950s negatively evaluate Gavin's anachronistic insistence. In *William Faulkner: A Critical Study* (1952), Howe comments: "When Gavin Stevens launches his lectures in behalf of Southern separatism and 'homogeneity' (a term happily left undefined), he becomes one of Faulkner's more tedious characters."[12] For Howe, Stevens's elocution in this novel is an echo of the South's common cliché. In *The Tangled Fire of William Faulkner* (1954), O'Connor agrees with Howe's opinion when he argues that *Intruder in the Dust* "sentimentalizes a region" and "argues a sectional thesis." O'Connor considers Gavin a person who "explains to Chick that the South is the only relatively homogeneous region left in the United States, and the consequent importance of maintaining this homogeneity," and concludes that this novel does not fully face the complex condition of the American South.[13] In *The Novels of William Faulkner* (1959), an anti-Southern critic, Olga W. Vickery, finds that Gavin's voice represents the self-justification of Southern whites' mass psychology.[14] Vickery argues that the Southern white community's homogeneity is expressed in the individual members' "complete loss of identity as well as of the power to act and think independently" in its "enactment of the ritual of lynching," and Gavin is "simply the most articulate spokesman for this mass reaction."[15] These critics presume Gavin is the mouthpiece of the Southern whites' stubborn sense of homogeneity, suggesting that this novel's theme is confined to a problem of the South's stereotype.

The following main critics in the 1960s do not question these early critic's presuppositions that *Intruder in the Dust* is regional. These critics argue the white community's unifying tradition is still living there, viewing Gavin's voice as representative of the white elites. Cleanth Brooks and Michael Millgate believe that Gavin's voice represents the Southern community's conscience to solve this problem by themselves. In *William Faulkner: The Yoknapatawpha Country* (1963), Brooks considers that Gavin's voice affirms the possibility of the unified white community's moral improvement, and concludes: "Gavin evidently finds in the unanimity of the community something of hope."[16] Millgate's *The Achievement of William Faulkner* (1965) also interprets that Gavin is the leading voice of the community but insists on this novel's sectionalism. Focusing on Gavin's "sermons on the South, its essential community, its need and ability to solve its own problems in its own way and its own time," Millgate argues *Intruder in the Dust* is a propaganda novel in which Gavin claims the local white community's independent power and makes Chick its agent.[17] As if to affirm these critics' readings, in his conversation with Chick, Gavin concludes: "I only say that the injustice is ours, the South's. We must expiate and abolish it ourselves, alone and without help nor even (with thanks) advice" (ID 199). If Gavin's opinion represents the main theme of this novel, readers should take this novel as a narrow regional work that encloses its racial problem inside the local agony. Gavin does not hear Lucas's words and insists that the white elites owe the abolition of racial injustice "to Lucas whether he wants it or not" (ID 199). In Brooks's and Millgate's arguments, the white character's voices and actions are intentionally displayed to proudly show the local white community's moral power to solve the racial problem.

Whether they like it or not, Howe, O'Connor, Vickery, Brooks, and Millgate find the Southern white community keeps its homogeneity in Faulkner's work. These approaches imply that this novel affirms the social privilege of the white Southerners who propagate their position after their confrontation with the budding civil rights movement. Critical studies in the 1970s inherit these early critics' attitudes when they claim that this novel is propaganda, identifying Gavin's voice with the author's. In *William Faulkner of Yoknapatawpha County* (1973), Lewis Leary concludes that "*Intruder in the Dust* is finally more effective as propaganda than as art" because "Faulkner seems to plead too hard for an understanding of the complex inherited views of the intelligent white Southerner."[18] In *Class and Character in Faulkner's South* (1976), Myra Jehlen believes Gavin is a person "who like Faulkner was willing to accept integration once blacks became worthy of it," and Chick is a boy who embodies the spirit of good whites, "who bravely shoulders the white gentleman's burden."[19] According to these critics' opinions, it might be natural to conclude that, as Leary says, *Intruder in the Dust* is "the most regional of Faulkner's novels."[20] If the racial issue in *Intruder in the Dust* is a particularly regional matter, the solution to this problem cannot be shared with outsiders.

Critics in the 1980s continued to praise the white male Southerners' subjectivities and their complex structures in Faulkner's works, while Faulkner's interest in the characters' intersubjective relationships in *Intruder in the Dust* is not so focused. Faulkner is a cultural hero that represents the artist's independence, and the white male Southerners' socially alienated minds are symbols of modern individuals' agonies. This attitude gives strong influence to the selection of the Faulkner canon. In the beginning of his fine study, *Faulkner: The House Divided* (1983), Eric J. Sundquist compares Faulkner's middle works (ex. *The Sound and the Fury*, *Absalom, Absalom!*, and *Light in August*) with his later works (ex. *Intruder in the Dust*, *A Fable*, and *The Reivers*) to suggest that Faulkner's later works are not his first-rate masterpieces. Sundquist argues that "Faulkner's deserved public recognition came at a time (the late 1940s and on to his death) when his best work was a decade old and he was writing some of the most disappointing fiction a major novelist could conceivably write" and that little "he produced after *Go Down, Moses*, including *Intruder in the Dust*, the novel that guaranteed that recognition." Sundquist's conclusion is that *The Sound and the Fury* is the mythic origin of Faulkner's works, because it contains "the repressed that returns with increasing visibility over the course of Faulkner's career as he discovers the lost dimension of Southern experience."[21] He treats *The Sound and Fury* as Faulkner's "myth of tortured innocence"[22] that occupies the central position where Faulkner's whole fiction will return again. Quoting John T. Irwin's impressive analysis on *The Sound and the Fury*, Sundquist explains why readers can assume Faulkner's South as the special place to which they are bonded and attracted:

> As Irwin points out, Quentin's attempt "to avenge his sister's lost virginity (proving thereby that it had once existed) and maintain the family honor is an attempt to maintain . . . the possibility of the existence of a virgin space within which one can still be first, within which one can have authority through originality." But the most tormenting denial of that virgin space . . . lies in the fact that "the Civil War closed off the virgin space and the time of origins, so that the antebellum South became in the minds of postwar Southerners that debilitating 'golden age and lost world' in comparison with which the present is inadequate."[23]

In Quentin's incestuous desire, the antebellum South is connected to an illusion of his sister's virgin space, in which one can keep his own privilege as the first one for the world. Because of its defeat in the Civil War, the South paradoxically keeps this illusion of the new genesis in the past, in the antebellum days. This lost world is still attractive for the modern individual who is frustrated in the modern world and wants to be the origin of a new space.

In contrast to the critics' high evaluations of the explorations of individual minds in his early and middle works, Faulkner's later works have been considered a reflection of the decline of his creative power. As Karl F. Zender argues in *The Crossing of the Ways* (1989), Faulkner himself presupposed that after World War II he would enter "the public arena and begin to function as a teacher and sage" in the world. In this process, Faulkner fundamentally changed his attitude on his works. Zender calls this period Faulkner's "mid-life crisis." In spite of his nostalgic affection for the good old South, Faulkner should revise his own imagined community, Yoknapatawpha County, for the sake of his current world readers. As Gavin's frantic eloquence shows, the Southerner's homogeneous identity with the Southern white community is in danger when their race problem is recognized in the outer world. According to Zender, Faulkner's *Intruder in the Dust* suggests his urge toward his new creative power, though he encountered two interrelated obstacles: "On the one hand, it was impeded by his reluctance to invest the scene of instruction with its full psychosexual complexity; on the other hand, it was impeded by the conflict between his desire to teach a liberal, progressive ethical lesson and his desire to hold on to the Yoknapatawpha County of his youth." The psychological and sexual complexity of white male Southerners does not coordinate with the universal theme of a people's common ethic. Zender considers that Faulkner's other novels of the late 1940s and early 1950s show similar irresolution: "the novels of this period focus simultaneously on instruction and on the impulse to revolt against it."[24] Zender's reading suggests that Faulkner's later work essentially shows serious self-doubt about his own world—that it now needs the moralistic answers to respond to the broad reader's demands.

These negative criticisms of *Intruder in the Dust*, however, overlook its attractive aspect. In this novel, the young protagonist's focus on his intersubjective relationships with other socially marginalized people claims to see the community not as the result of decent people's pursuit for a self-sufficient society but as the result of multiple connections of people who have problems that are "not in solution but in suspension" (ID 49). These connections do not give any secured positions to a particular social group, nor do they give any catastrophes to particular victims' tragic endings. Developing intersubjective relationships with others is a young white protagonist's way to recognize the world as the process of people's "agonies and shames and griefs" (ID 49). In *Intruder in the Dust*, the characters' words—which show their own powerless, marginalized positions—are effective to create the space in which any utterances cannot gain complete dominance. Shifting the focus from subjectivity to intersubjectivity can be considered Faulkner's will to create a new imagined space for the world.

Studies in the 1980s almost complete a Freudian/Lacanian psychological analysis of Faulkner's works. Scholars believe Faulkner's greatest middle works are an analyses of the complex psychological self of the Southern white male. The main characters are Quentin Compson, Joe Christmas, Thomas Sutpen, and Ike McCaslin.[25] These characters are obsessed with the ideal self in the traditional South, and they want to eternally identify themselves with the white persona. This desire also assumes that the eternal continuation of the same kind of community keeps the same value in its core. Whether they appreciate them or criticize them, the main studies of Faulkner's works (from the 1940s to the 1980s) focus on the continuation of the same selfhood of the elite white male characters that believe in a monolithic social core in the American South.

After the flourishing of psychological analyses on Faulkner's characters, a study of the cultural formation of Faulkner's works began near the end of the 1990s. As many scholars have already pointed out,[26] Michael Kreyling's *Inventing Southern Literature* (1998) is one of the earliest studies to remap Southern literature as a cultural product, and to criticize the literary tradition of the American South as some local elites' intentionally produced images. One of Kreyling's main targets is Faulkner's works—Faulkner is now a cultural icon of the American South, and "if the 'South' is a cultural entity, then 'Faulkner' is its official language." Kreyling studies how critics have chosen Faulkner as the American South's representative figure, and how they create an author as an "institutionalized cultural force." Following Benedict Anderson's argument on imagined community, Kreyling considers that community is fictively created with various people's contemporary cultural acts, and that identity is "not an innate phenomenon but a product culturally and historically fabricated to local specifications by narratives that are more or less cooperative . . . and more or less conscious."[27] Kreyling read Faulkner's later works to remap Faulkner again in history and examine how Faulkner, especially after winning a Nobel Prize, reluctantly dealt with his fame and his reputation. He also examines how critics have invented the South and opens up how Faulkner's works have mythically represented the Southern history. After Kreyling, two decades worth of studies of the American South and its culture—which include Limón's *American Encounters* (1998), Gray's *Southern Aberrations* (2000), Yaeger's *Dirt and Desire* (2000), Baker's *Turning South Again* (2001), Jones and Monteith's *South to a New Place* (2002), Hobson's *South to the Future* (2002), McPherson's *Reconstructing Dixie* (2003), Smith and Cohn's *Look Away!* (2004), Winchell's *Reinventing the South* (2006), and Schmidt's *Sitting in Darkness* (2008)—criticize the "mythifying energy of Faulkner studies,"[28] which reproduce Faulkner's works as the American South's fundamental cultural asset, in order to demystify Faulkner's social position in his contemporary world and remap Faulkner's works in the process of the region's cultural formations.

Jon Smith's *Finding Purple America* (2013) is one of the more recent studies to reconsider Faulkner's image as performative in today's world. For Smith, Faulkner is "a good deal hipper than his critics have been." Faulkner's works not only radically fight against the cultural mainstream but also function as self-conscious improvisations for current movements of the world. Unlike an ancient painter of Greek urns, Faulkner does not have a highly wrought design for his idea from the beginning, but his works are written to contain unstable concepts that respond to current cultural formations. Smith calls Faulkner's writing a "fashion," meaning Faulkner's language is "symbolizing social mobility and change." It floats between, not within, social groups. Focusing on one of Faulkner's later works, *The Mansion*, Smith insists that Faulkner uses his language to connect the resistances of "various kinds of oppressed men" with his own "miserable" resistance against the "main-stream America's 'intolerance.'"[29] Smith considers that what is focused on in Faulkner's expressions is not the subject that is deeply rooted in one culture, but the language that connects the author's position with marginalized groups.

Kreyling and Smith focus on Faulkner's later works to reconsider Faulkner not as a mythic Yoknapatawpha kingdom's author but as a part of the cultural formation in the middle of the twentieth century. To develop their arguments, this study focuses on *Intruder in the Dust* as Faulkner's turning point from his early and middle works' explorations of individuals' subjectivity to his later works' analyses on intersubjective relationships in a socially stratified community.[30] This novel's intersubjective connections between characters suggest the alternative processes of a community's formation. The South is not the place for a monolithic community's continuation, but the place for a heterogeneous people's multiple connections.

Focusing on the personal communications in deeply stratified societies allows me to connect Faulkner's works with other literary works of the American South. For example, Chick's discovery of the value of intersubjectivity among the socially marginalized individuals in *Intruder in the Dust* relates to the conclusion of Toni Morrison's *Beloved* (1987). In Morrison's novel, Paul D persuades Sethe to evaluate their private relationship and reconsider their future: "Sethe . . . me and you, we got more yesterday than anybody. We need some kind of tomorrow." He declares that Sethe can love herself as her own "best thing." Sethe rediscovers her own value in her intersubjective relationship with Paul D, exclaiming "Me? Me?"—to reestablish her own position, Sethe sees herself from the alternative point of view that she creates with Paul D.[31] The climax of Morrison's novel resonates with Chick's establishment of his personal relationship with Lucas in *Intruder in the Dust*. Another example is Valerie Martin's novel, *Property* (2003), which focuses on the antebellum South's intersubjective relationship between a planter's wife and a slave woman. The planter's wife knows the slave is her husband's lover, and under the threat of slave rebellion they form a different kind of relationship. These authors' continuous efforts to explore and evaluate people's personal combinations in every social form suggest another context of literary history of the American South.[32]

I use social combination as a term that contrasts with what Hagood calls "an imperial impulse" in Faulkner's text. In his study, *Faulkner's Imperialism*, Hagood argues that Faulkner's Yoknapatawpha County is a mythic, imperial space that insists on the importance of "maintaining stability and power." The "mythic-imperial layers," which can be traced back to the imperial power depicted in "Greco-Roman, Egyptian and Arthurian myths," are the keystone of Faulkner's cosmos. A typical model of this mythic Southern space appears as a metaphorical parallel "between the Mississippi Delta and the biblical Egypt" in *Go Down, Moses*. Hagood argues that in *Go Down, Moses* these socially stratified people—African Americans at the plantations, Native Americans in the Delta, and the white people who are imprisoned by the curse of their heritage—are all enslaved in the imperial South and demand to be free from their "mutual enslavement." Faulkner's white male elite characters do not enjoy their position because they are also slaves of the imperial layer of the South. Their consciences are threatened by the oppressed voices of marginalized groups such as women, Native Americans, poor whites, and African Americans. This condition suggests the importance of these marginalized groups' relationships with the white male characters. Although his argument is quite persuasive, Hagood—who focuses on the various layers underneath imperial design in a culture—does not give a particular name for the marginalized people's connections that create their "narratives of subversion" to resist against the imperial dominance. Hagood treats the voice of resistance as marginal one because it only occurs when it confronts imperial will, assuming that humans' inexhaustible voice essentially comes from "the mythic place and imperial space of Faulkner's cosmos." Therefore, if readers focus on individual characters' voices, it seems Faulkner's characters are always obsessed with the anachronistic, imperial desire of the mythic South to govern the people who live there.[33]

Hagood correctly evaluates *Intruder in the Dust* as a novel that can give a more hopeful outcome in the community. He argues, "[Chick] Mallison might be the white man who can carry southern society a step farther even than Ike McCaslin by joining forces on more equal terms with women and African Americans." Supposing Chick in the South is Moses in Egypt, Hagood predicts that Chick "might be the leader of a new dawn for Yoknapatawpha race relations."[34] His argument, however, does not explain in what way a sixteen-year-old boy can be temporarily free from the American South's profound gravity of imperial desire to govern a socially marginalized people. If it is true that imperial will is rooted in the origin of human culture, the young protagonist needs to find an alternative value that can persuasively show another phase of the world in his own struggles with the dominant discourse of the community. How can Chick provisionally escape from the community's dominant imperial impulse and its preference for lynching? Moses escaped from Egypt and guided his people to the promised land because of his discovery of a new god. Chick must also find another value in order to discover an alternate layer in his community.[35] In what paradigm can Chick arrive with the other community members? Hagood's argument does not answer this question.

What Chick finds in this novel is a temporal opportunity for dialogical intersections between individuals under the pressure of white dominance and violence. Chick is sensitive enough to recognize how his conversations with the other socially marginalized characters are undermining his own imperial impulse. In spite of the supreme power of the white male adults' imagined community, the culture of the American South has kept personal space for democratic dialogues between socially marginalized characters beyond the border of race, gender, and class. In this alternative social structure, even if participants of dialogue demand to preserve their own powers in conversation, they find that their personally shared words reduce each dialoguer's dominant power. Dialoguers use their language to communicate with others in order to know their language gains meaning not only from their own will, but also from what they create with others. What readers find in *Intruder in the Dust* is this layer of language that temporarily discharges characters from their imperial impulses. Every individual's effort to communicate with someone else rebels against the desire for self-preservation and self-unification.

In their relationships with others, individuals feel that others' words are inevitably included in their conscious. Bakhtin argues that this innate desire of language to merge together is a fundamental sensation that struggles with the metaphysical idea of the unified self. It is the living sense of merging with others:

> *Not-I* in me, that is, existence in me; something larger than me in me. . . . I enter into the spatial world, but the other has always resided in it. The differences between space and time of *I* and *other*. They exist in living sensation, but abstract thought erases them. Thought creates a unified, general world of man, irrespective of *I* and *other*. In primitive, natural self-sensation, *I* and *other* merge. There is neither egoism nor altruism here. (FNM 146–47)

Bakhtin assumes the common sensation between *I* and the *other* undermines the participants' totalized selves in order to claim the alternative phase of human consciousness. Not only people's imperial impulses, but also the mutual intercourses individuals have with others, are primitive and ancient. The living sensation of merging the self and another has been diminished and ignored by the modern, Westernized idea of a unified self, but it still exists in our everyday experiences.

According to Bakhtin, sticking to the totalized self and its dominant power shows the narrowness of the Western world's vernacular in the nineteenth century. Bakhtin claims ideal literature includes a boundless variety of voices and celebrates their relationships. In the following argument, when he says "the East," Bakhtin includes everything that is not considered enlightened Western culture in the nineteenth century:

> Science (and cultural consciousness) of the nineteenth century singled out only a miniature world (and we have narrowed it even more) from the boundless world of literature. This miniature world included almost nothing of the East. The world of culture and literature is essentially as boundless as the universe. We are speaking not about its geographical breadth (this is limited), but about its semantic depths, which are as bottomless as the depths of matter. The infinite diversity of interpretations, images, figurative semantic combinations, materials and their interpretations, and so forth. We have narrowed it terribly by selecting and by modernizing what has been selected. We impoverish the past and do not enrich ourselves. We are suffocating in the captivity of narrow and homogeneous interpretations. (FNM 140)

The narrowness of nineteenth-century Western culture and literature is contrasted with the primordial broadness and diversity of language. The intersubjective layer of literature is suffocated in the nineteenth century, but Bakhtin's argument aims to celebrate infinite varieties of literature in order to break through the limitations of the homogeneous interpretation of the world.

The nineteenth century's narrow idea of a unified self and community still attracts those who demand to keep their dominant power in the world, even if it also finds the world's language realizes its diversity. Bakhtin uses several expressions to focus on this dynamism that move in two different directions. On the one hand, the people's metaphorical idea is attracted to the homogeneous unifications of their selves and their monolithic community to create a "concrete verbal and ideological unification" (DN 271); on the other hand, people's actual languages spread to the heterogeneous combinations of their selves and others in cultural multiplicity, realizing the power of "stratification and heteroglossia widen and deepen as long as language is alive and developing" (DN 272). Bakhtin calls the former "the centripetal forces" and the latter "the centrifugal forces" of language (DN 272).

Enlightened Western culture demands to build a stabilized rationale to dominate their perpetual community using their language's centripetal forces; however, when a free and independent individual has a conversation with another, the primitive merge of the self and the other occurs as an innate phenomenon of the language's centrifugal forces. What Faulkner shows in *Intruder in the Dust* is temporal appearances of this intersubjective layer of language, which the young protagonist feels in his communications with the marginalized members of the community. Modern individuals feel that the whisper of their egalitarian impulse comes from their language's intertwined condition with others. This intersubjective layer, which cannot be dominated by the idea of the individually unified self, guarantees the community's fundamental instability and multiplicity. Marginalized characters' conversations indicate this subversive layer below the community's white male imperial dominance.

Both Faulkner and Bakhtin are not interested in the official procedure for realizing the democratic politics in a community. Faulkner believes the racial problem is not a matter of the federal government, but it "belongs to the South" (LG 60). In his essay titled "On Privacy" (1955), Faulkner argues that equality is a fundamental, interpersonal responsibility that has no relationship with politics: people possess "liberty in which to have an equal start at equality with all other men, and freedom in which to defend and preserve that equality by means of the individual courage and the honorable work and the mutual responsibility."[36] Faulkner assumes that only personal actions can reaffirm innate equalities. Similarly, Bakhtin, whose argument is demonstrated in Dostoevsky's novels, shows a tendency to disregard the political systems that formally govern people's relations. Bakhtin focuses on individuals' communicative dimensions in people's words, which can be a common plane on which individuals can be "juxtaposed dialogically" (DN 293). Even if their society is stratified or classified, characters' various interests toward each other betray their demand for mutually shared sensations. Bakhtin's attitude toward the diversity of people's voices and their innate tendencies to create common meanings suggests that his arguments ultimately aim for the common layers in people's personal relationships. As Hirschkop suggests, Bakhtin's argument indicates that "democracy and equal rights are not law but language, not the product of institutions above society but of the linguistic intersubjectivity which constitutes it."[37] In the works of both Faulkner and Bakhtin, a democratic culture is imagined in people's interpersonal activities. Faulkner's *Intruder in the Dust* tells a story in which the local characters develop their abilities to have democratic relationships using their vocal exchanges, just as Bakhtin's argument supposes the ideal feature of the novel in the democratic structure of language.

After World War II, Faulkner, who joins many intellectuals and artists in imagining how a nation can liberate itself from the burden of racism, intentionally gives a tendency toward "a genuine post-racial mentality" to his characters.[38] In this context, a white boy's encounters with African Americans, a white spinster, and a poor white family can be read as symbolic confrontations that suggest the broader world's social unsteadiness,[39] and Gavin, the county attorney who insists on the honorable consistency of the Southern white community, can be interpreted as the idealistic theorist who believes in the dignity of individualism and reductive racial categories. What Chick finds in this novel is the way to perceive and refer to personal common grounds, while the socially divided people keep their instabilities in their community.

Temporal efforts to expose the common layers in their personal communications appear in *Intruder in the Dust*. As his controversial sentiments on the racial problem are distributed and refracted in his characters' minds, Faulkner does not allow anyone to have a monotonous speech. As Faulkner's letter suggests, what he wants in *Intruder in the Dust* is "a story," and he believes "nobody preaches in it" (SL 262). Nobody, including the author, can give the total perspective on a racial matter. Their words indicate the possibility of equal, but incomplete, relationships in the community. A common layer in interpersonal relationships is temporarily suggested in people's social combinations, but this premonition soon disappears again. The detective process to discover the serial murderer relates to the characters' desire to show their mutual trust based on their democratic sensations—Faulkner's socially marginalized characters cannot find any dominant language that expresses their relationships. Moreover, this novel suggests how the white male adults' imperial impulses and their authoritative words disregard and oppress the socially marginalized people's personal relations. This novel concludes by discovering the ways a character can evaluate the democratic possibility of personal relationships with others in a deeply stratified society.

The following three arguments examine how this novel's protagonist, Chick, develops his social combinations with the marginalized characters.

The first part inspects the way Chick finds the intersubjective relationship with Lucas in order to cope with the uncertain situation of Lucas's lynching. In the second part, Chick's communications with Aleck Sander and Miss Habersham are read as a young protagonist's process for evaluating his temporal communications with socially marginalized people. In the third part, the social formation of the local community is analyzed, along with Chick's egalitarian worldview and decision to resist the logic of Gavin, a white lawyer who develops eloquent logical arguments to keep his position as the ideal guardian of the African American. I argue that Chick's recognition of the world is a metaphor for today's individuals who are connected to others by digital networks. The development of Chick's worldview not only reflects the expansion of a young Southerner's relationships with other locals, but it also reflects the modern people's dialogical condition to expand their mutual links. Chick, who connects himself with the marginalized people, symbolizes not only the American South's budding civil rights age but also our contemporary hopes to create the next social formation by our expanded connections with others.

CHICK'S SOCIAL COMBINATION WITH LUCAS

This novel's communication between Chick and Lucas suggests that, although the white male adults' dominant discourse impedes socially marginalized people's ability to have personal relationships, these socially marginalized people can still find ways to share their own intersubjective recognitions in equal conditions. *Intruder in the Dust* begins with two characters' intersections beyond the racial border. On the one hand, Lucas steps over the color line because of his arrogant attitude toward whites. He needs to communicate with Chick under the threat of lynching, and he expects that his own shortcomings regarding the language will be filled by his intersubjective relationship with Chick. On the other hand, Chick tries to temporarily see the world from Lucas's viewpoint in order to avoid his white Southerner's pride that makes him demand lynching, violence, and death. Lucas persuades Chick because Lucas's words are filled with hope for life instead of death.

Chick's social combination with Lucas begins with their effort to interpret the nonverbal sensations they share. One of those nonverbal elements is the food Chick eats at Lucas's dwelling. In the beginning of their relationship, Chick is a boy of twelve who goes rabbit hunting with Aleck Sander, his African American childhood friend, and Joe, a young African American guide. He accidentally falls into the icy river, but Lucas walks by and gives advice that helps him escape from the water. Lucas, a proud black farmer whose grandfather was a white planter in the antebellum period, brings the cold and wet Chick to his house. Chick temporarily shares the same living items with Lucas. While Chick is drying his wet clothes at Lucas's house, Chick uses Lucas's quilt and feels he is enveloped in the "unmistakable odor of Negroes" (ID 11). Moreover, Chick eats half of Lucas's dinner—what they call "nigger food" (ID 13). It is typical of the food African Americans eat for dinner, but Southern whites do not normally eat in an African American's house, nor do they use African American dishes and utensils.[40] Sharing the same "metabolism" (ID 13) with Lucas, Chick senses that the meal has bound himself and Lucas in an organic fashion. Chick feels this primitive relation with Lucas is humiliating, and his white Southerner's pride demands to certify his superior position over Lucas. Chick tries to pay Lucas seventy cents for his treatment, but Lucas does not take the money. Because of his pride in being an African American descendant of the one of the oldest white families in the area, Lucas took care of Chick as a mutual member of the old lineage. He calmly rejects the coins Chick offers. Chick drops the coins on the floor and orders Lucas to pick them up, but Lucas stays still and again orders Aleck Sander and Joe to pick up the coins and return them to Chick.

In this early stage of their communication, because both Chick and Lucas want to keep their own proud positions, they do not know the proper manner for evaluating their temporal relationship and their shared living sensation. After returning home, as a reward for Lucas's hospitality, Chick sends a Christmas present: "the four two-for-a-quarter cigars for Lucas and the tumbler of snuff for his wife" (ID 22). But Lucas gives "a gallon bucket of fresh homemade sorghum molasses" (ID 23) in return. During the four years after their encounter, Lucas skillfully keeps his advantageous position over Chick. The memory of rejected payment and a sense of debt haunt Chick in the form of an enlarged image of a coin like "the dead monstrous heatless disc" that hangs "nightly in the black abyss of the rage and impotence" (ID 22). Since he cannot control his relationship with Lucas, Chick is threatened by a sense of "impotence" as a white male, and he feels his subordination to Lucas as if he had fallen into "the black abyss." Chick exaggerates the sensation he shared with Lucas as a fatal matter for his self. His identity as a white Southerner demands a separate, distinguished place where he can keep the power of supremacy against the African Americans, and he believes what he shared with Lucas is a fatal error that threatens his pride.

Chick's desire to keep his superior position above Lucas increases to homicidal thoughts, with the help of the dominant discourse of the white males in the community. Although Lucas is categorized as an African American, the costume he inherited from his white grandfather suggests that one of his ancestors was a member of the oldest white families in the area. Lucas appears in town with his white grandfather's antique costume and pistol every Saturday night. In a grocery store, Lucas proudly says he belongs "to the old lot" (ID 19), and uses his blood connection with the old white family to look down on the other common white fellows. Understanding what "every white man in that whole section of the country" has been thinking about Lucas for years (ID 18), Chick senses the white people's malice against Lucas. The town's white men are almost all irritated with Lucas's aloofness. He hears a white man in a store curse Lucas as a "goddamn biggity stiffnecked stinking burrheaded . . . sonofabitch" (ID 19). Internalizing the white community's vulgarity, Chick wishes Lucas "*would just be a nigger first, just for one second, one little infinitesimal second*" (ID 22). The lowly invective of Chick's community supports a white boy's acute desire to despise Lucas. Chick's internal malice against Lucas is supported by the town's dominant discourse.

Chick's struggle with Lucas continues in his internal dialogue. When he accidentally meets Lucas in the center of Jefferson three years after their first encounter, Chick finds that Lucas does not recognize him. Chick has already heard that Lucas's wife died. Because he is deeply mourning his wife, Lucas disregards any current matters when he walks around Jefferson. Lucas has a chance to see Chick's eyes directly, but his countenance does not show any recognition. Chick says he feels "free" because the man he has been obsessed with "for three years" has "walked out of" him (ID 26). Chick finds a way to separate his relationship with Lucas and recognize his own self as complete and independent. This is not his "revenge" against Lucas, but his demand for "re-equalization, reaffirmation of his masculinity and his white blood" (ID 26). This sense of equality is not demanding to communicate with Lucas again, but to establish individuality in his relationship. Chick's ideal relation with Lucas is mutually independent, allowing him to keep his own pride without remembering his humiliated past. Therefore, he imagines himself and Lucas as old men who cannot remember their past, and their mutual humiliations are forgotten as meaningless matters for individuals who live separately. Chick could imagine himself and Lucas as old men meeting, and

> he saying to Lucas: *I was the boy who when you gave me half of your dinner tried to pay you with some things which people in those days called seventy cents' worth of money and so all I could think of to save my face was to fling it on the floor? Dont you remember?* and Lucas: *Was that me?* or vice versa, turned around and it was Lucas saying *I was the man when you throwed your money on the floor and wouldn't pick it up I had to have two niggers pick it up and hand it back to you? Dont you remember?* and he this time: *Was that me?* Because it was over now. He had turned the other cheek and it had been accepted. He was free. (ID 26–27)

His internal conversation with Lucas secures his desire to eternalize their separation and be free. This internal imagination also means that their mutually independent positions demolish their vitalities. Both Lucas and Chick are old and near to death, which will eternally separate them. Chick's imagined conversation with Lucas shows a limitation of Chick's idea of mutual independence in a community. His independent self cannot find a way to enliven his life and evaluate what he shared with Lucas.

Chick's obsessive demand for the preservation of the individualized self and for disconnecting his relationship with Lucas is a common desire in Faulkner's white protagonists who inherit the Southern idealized image of the self. As Richard C. Moreland suggests in *Faulkner and Modernism* (1990), the white male protagonist's ideal self appears in "his internalization of his society's idealized image of the planter."[41] Faulkner repeats and revises the antebellum South's self-content planter image as a typical type of self that urges the white male elite's demand for the independence. The protagonist of *The Sound and the Fury*, Quentin, shows this pattern's ultimate conclusion. He kills himself because his society's pressure for the preservation of identity is stronger than his survival instinct. As Quentin's suicide shows, the modern Southerner's desire to maintain an independent, unified self can become equated with a violent passion to demolish another desire to share the living sensations with the others.

In *The Sound and the Fury*, Faulkner's modernist technique works well to critically depict the modern Southerner's obsession with the static preservation of the ideal world as an inevitable malady of the twentieth-century's Euro-American culture. Modern, Westernized individuals have a strong desire to assume their independent, eternal positions, and they suppress the fact that their transient connections with others cannot allow them to have eternal positions in their community. In *Faulkner's Subject* (1992), Philip M. Weinstein criticizes Quentin's obsessive demand for his ideal world's independence as a typical element of the Euro-American identity, to suggest it is the main cause that demolishes Quentin's will to tolerate his life being involved with other people's voices:

> I suggest that identity is a privileged term within a Western vocabulary of individualism. In its primary meaning, that something is always itself, and its secondary implication, that that selfhood is different from all others, the concept of identity makes some large promises. It promises sameness over time, an unchanging essence at our core. Our self-sameness is intrinsic (to be found within) rather than relational (to be found by way of our membership within a larger group). The term suggests, further, that we are unique creatures, essentially different from each other. To privilege individual identity in this way is to fantasize a protected sacred place, the place of ourselves, which would be immune to the vicissitudes of time, space, and culture. It is to allay our anxiety that we may not have an unchanging core and therefore may take on our meanings from our affiliations and conditions. It is to fix, enclose, and affirm our unique difference from others, to say: "That's who I am." By thus reifying our

> sense of ourselves, by charting it as a separable essence and putting boundaries around it, we repress that intolerable sense of *being-helplessly-caught-up-in-the-Other* that Faulkner represents in the plight of Quentin Compson.[42]

The Euro-American demand for the continuation of an unchanging self has a controversial relationship with the larger society's modern development and local variety. Though a self may want to be identified with the eternal sameness in an imagined monolithic community, modern enlarged society, which requires the self's transformation according to the outer changing circumstances, does not appreciate the self's stubborn identity. People's social lives change in response to time, space, culture, and the fluctuations of interpersonal relationships with others threaten the self's eternal sameness. Quentin's incestuous desire for his sister and his suicidal attitude is connected with his demand for the eternal preservation of his own internal world, and it also suffocates Quentin's other aspect which openly accepts the other people's voices. If a self ultimately wishes to preserve its own identity, it requires the total eradication of the living sensation that can be shared with others.

When Chick demands the eternal detachment of himself from Lucas in his imagination, Chick follows Faulknerian character's typical psychological obsession with self-sameness. This immoderate, self-centered passion has only two goals: the subject's own self-destruction or the destruction of others. In Quentin's case, he chooses to kill himself, but Chick's demand to preserve his self does not increase his suicidal urge—instead it resonates with the white community's sadistic malice toward the old African American. When he hears the murder case of Vinson Gowerie and the possible lynching of Lucas, Chick's first reaction is to expect Lucas to look back at him during the lynching's "gasoline-roar" (ID 41). To justify this cruel desire, he tries to believe he is "free" from Lucas and that Lucas is "no longer his responsibility" (ID 41). To excuse his sadistic attitude toward Lucas, he insists on his independence and freedom from Lucas. His living sensation, which still tells of his personal connection with Lucas, is repressed when Chick sticks to his independent freedom.

Although his independence from Lucas is completed in his damnatory imagination, Chick's nonverbal sensation still recognizes their relationship. Chick's obsessive feelings about Lucas bring him to the street in front of the jail. It is reported that Lucas was found standing over the body of a young, poor white, Vinson Gowrie, with a fired pistol in his pocket. In spite of a decrease in frequency, the fear of lynching still smolders in the community at the time.[43] People expect that Vinson Gowrie's family members will come to lynch Lucas at night. In spite of his desire to be unrelated to Lucas, Chick still demands Lucas's recognition of him. When Lucas arrives "in front of the jail" (ID 42), Chick is there as a white boy who knows Lucas "as well . . . as any white person" (ID 3–4). As if to agree with his four-year obsession, Lucas identifies him. "Looking straight at him," Lucas addresses him: "You, young man" (ID 44). His gaze reawakens Chick's personal connection with Lucas. When Lucas says, "Tell your uncle I wants to see him" (ID 44), Chick understands that Lucas recognizes him and wants him to bring Gavin to the jail to be his lawyer.

Chick's communication with Lucas makes him the only white community member to establish a personal relationship with Lucas. Lucas partly claims this intimacy with Chick, relying on their debtor-creditor relationship caused by their shared food four years ago, and at the same time he appeals to Chick using nonverbal demands. In the middle of the mob's threat of lynching, Lucas's concurrent demand for the solution has strong appeal to Chick. Chick feels that Lucas's personal desire for life is more persuasive than the mob's dominant discourse to nullify Lucas's existence for the community's self-satisfaction. Lucas's persuasiveness comes from Chick's hateful recognition of his identification with the dominant discourse. After Chick decides to accept Lucas's demand, Chick began to perceive that his world as the "living entity at this point at this moment on this day" (ID 46). Here Lucas and Chick create their social combination, by which they can make sense of their shared sensation for life in their contemporary world.

Simply put, as Millgate argues, Chick tries to stop the lynching "in favour of a positive and even passionate awareness of the need to preserve human dignity and avert the shame of mob-violence."[44] This human dignity gains real meaning when Chick recognizes the universality of a man's desire for intersubjective connections rather than hostility between socially stratified groups. The utterance of the old African American gains internal persuasiveness because his request represents essentially the world's controversial requirement: the world is deeply stratified, but shared living sensations among individuals also demand common ground in order to connect individuals. As Bakhtin argues, "The internally persuasive word is either a contemporary word, born in a zone of contact with unresolved contemporaneity, or else it is a word that has been reclaimed for contemporaneity" (DN 346). As racial problems in a community are part of the problems of a stratified society, it is meaningful to consider this localized race issue as an analogy for the world's unsolved problem. Chick perceives his community's malice based on his own personal connections with Lucas, but this structure suggests their actions reflect the world's effort to fight the social forces of stratification.

The dominant voice in Chick's world demands lynching and social stratification, but personal voices create alternative meaning and context in the socially marginalized members' relationship. Generally, the cultural pattern of the American South at the time, as Donna Gerstenberger says, still allows "the hanging of a Negro," if that person revolts against the whites.[45] In order to justify the social mores, the whites must presuppose African Americans as potential villains. However, upon confronting Lucas's predicament, Chick discovers his imperial impulse has taken part in the mob's prejudice. Understanding Lucas's desire for life, Chick wants to avoid his own desire to lynch African Americans. To criticize his own inner cruelty as "something shocking and shameful out of the whole white foundation of the county" (ID 135), Chick chooses to take part in the intersubjective connection with Lucas's worldview.

Chick, now aware that his internal logic as a white community member does not give any solutions to this issue but instead endorses unlawful violent urges against African Americans, tries to hear Lucas's words in order to know Lucas's vision. The economy of Lucas's words enforces this relationship effectively. When Lucas briefly says, "Go out there and look at him"(ID 66), Chick understands that it means Lucas wants him to exhume Vinson's body from the grave and check the bullet hole in order to prove Lucas's innocence. Lucas's trust in the intersubjective relationship with Chick appears in his shortage of words. Lucas does not clearly explain his idea in front of whites, even in his personal conversations. Some critics feel Lucas is "virtually silenced"[46] or "outside the realm of speech"[47] because Lucas's condition symbolizes the failed—or psychologically castrated—masculinity of Southern blacks. However, it might be more reasonable to consider that Lucas carefully uses the terse language to avoid white authority. Lucas uses vagueness to force Chick to listen to what he has to say. If it is possible to apply Bakhtin's terms to this condition, Lucas knows his language "exists in other people's mouths" (DN 294). With strategically open-ended suggestions, Lucas uses his own powerless position to introduce a common ground between him and Chick. It is Chick who should find the words to vocalize Lucas's intention.

Lucas and Chick's social combination creates an alternative way to use their words to resist the community's dominant ideology. In his oppressed condition, Lucas shows his own matured linguistic intentions when he gives enough hints to communicate with Chick. Chick hears what Lucas wants to say based on his experience of their mutually shared sensation, and he creates his own interpersonal relationship with Lucas in order to resist the ideological discourse of the dominant Southern society. This is an alternate way for Chick to be a matured figure. Quoting Bakhtin's distinction between "authoritative discourse" (DN 344) and "internally persuasive discourse" (DN 345), Weinstein argues that "our journey as subjects from the former to the latter—from dicta that have been pressed upon us to those that we can internalize and make our own—is, precisely, our ideological maturation." Weinstein only judges Chick's experience with this maturity while he supposes "Lucas undergoes no such linguistic trajectory" because "Lucas does not speak."[48] However, if we literally take Weinstein's argument on people's ideological maturation,

Chick's experience with Lucas seems to indicate that one's internally persuasive discourse comes not only from the utterance of words but also from our expectations of sharing nonverbal recognition with others. People often hear others' silenced intentions in pauses between their words. Lucas also matures in his communications with Chick when he recognizes his ability to control his own position using his nonverbal/verbal relationships.

Lucas, knowing that Chick understands his critical view of the white community, gives brief hints to Chick to visit Vinson's tomb at night and "violate the grave" (ID 93). Gerstenberger argues that Chick is "free of the trappings of society" in his mission, and "sees what must be seen (however complex and unsatisfactory) with his own eyes."[49] However, more precisely, Chick examines his community not only from his own eyes but also from Lucas's perspective. As Carl J. Dimitri writes, Chick has to question "the prevailing social codes and rituals" and adopt "a new perspective from which he views Lucas and black-white relations."[50] Chick creates a reciprocal view with Lucas, which implies that both black and white men should cooperate to gain a satisfactory understanding of the facts.

To hear and understand the intentions behind the other person's words, Chick and Lucas have to understand the intersubjective, living condition they create. This situation reminds Chick of an episode with Old Ephraim, Aleck Sander's grandfather. When he was eleven, his mother searched everywhere in the house to find her lost ring, but she couldn't find it. To help her search, Old Ephraim, an old African American who could gather plenty of information from a poor white woman and the other African Americans, gave Chick useful directions. Old Ephraim said Chick should search "under the hawg-trough at your pa's farm" (ID 69). Chick's father considered Old Ephraim's words to be untrustworthy, but Chick and his mother resisted the white gentleman's "furious repudiation," went out to the farm, and found the ring "under the trough where the hogs were fed" with the help of an overseer (ID 70). Old Ephraim's suggestions were vague but effective. As no explanation was provided on the precise location of the trough, Chick and his mother needed the help of the overseer in order to find the ring. Following the elderly African American's directions, they stepped outside the usual social norms and got what they actually wanted.

Finding particular meaning in ambiguous words is the privilege of marginalized people. As Wesley Morris and Barbara Alverson Morris say, "The dominant race, gender, class cannot listen to anything outside the familiar expectations, cannot say anything not already said."[51] This communicative inability of the ruling class's aloof self-fashioning is illustrated in *Intruder in the Dust*. In his advice to Chick, Old Ephraim suggests the marginalized members' communications can inspire a new solidarity to gain what they really need:

> Young folks and womens, they aint cluttered. They can listen. But a middle-year man like your paw and your uncle, they cant listen. They aint got time. They're too busy with facks. In fact, you mought bear this in yo mind; someday you mought need it. If you ever needs to get anything done outside the common run, dont waste yo time on the menfolks; get the womens and children to working at it. (ID 70)

Old Ephraim's trust in marginalized people's communicative abilities explains why Chick can understand Lucas's intentions. As Stephanie Li argues, Old Ephraim's "words of wisdom are remembered privately by Chick and represent a knowledge."[52] Similar to African Americans, women and children are also symbols of marginalized powerlessness. As they share a socially powerless position, they do not believe they have the decisive words internally, but they can hear other people's words and make the temporal agreement in their specific contexts. Old Ephraim's words, "they aint cluttered," refer to people's ability to create a new way in their dialogized words in order to gain what they want for their lives. Old Ephraim suggests this personal formation for their lives can be useful in seeing the world from another point of view.

Old Ephraim's words suggest that socially marginalized people can more easily escape their ordinary norms and routines to solve problems. For Old Ephraim, the white elites' pride and self-defensive logic are obstacles to create a dialogic space among individuals. As Gavin's attitude toward Lucas suggests, the white adult male cannot hear African Americans' demands because their ideological position is established by excluding African Americans from their discourse. As Masami Sugimori suggests, "the reductive function of racialized language prevents whites from thinking (and thus acting) outside the scope of ideology."[53] In contrast, hearing Lucas's vague words, Chick remembers his dialogue with Old Ephraim and recognizes what is persuasive for him is not the dominant ideology but the socially marginalized member's incessant demands for life.

Society is always stratified. Language is reproduced as a fundamentally diversified condition. As Bakhtin argues, "Oppositions between individuals are only surface upheavals of the untamed elements in social heteroglossia, surface manifestations of those elements that play *on* such individual oppositions, make them contradictory, saturate their consciousness and discourses with a more fundamental speech diversity" (DN 326). This socially stratified condition, however, produces personal demands for connections between these groups. Old Ephraim's words performatively show that his locally diversified words ("Young folks and womens") relate to his demand for personal connections and dialogues with other socially stratified members ("They can listen"). This means every language mediates interactions between diversified individuals and this function should be activated in our personal connections. "Lucas was not even asking him [Chick] to believe anything" (ID 71) because they are different, but Lucas activated language's fundamental demand for dialogue in his relationship with Chick.

Faulkner illustrates that Chick's maturity comes not from his obsessive demand for independence, but from his intersubjective experience with Lucas. Chick finds that the experience of the temporal equality among individuals gives deep effects to their deeds. In *The Dialogical Self*, referring to Bakhtin's idea of dialogue, Hermans and Kempen argue that, throughout the history of psychology and philosophy, the theory of decentralization of the self has begun to doubt the existence of "a highly centralized ego in full control of its own thoughts," and focusing on the social relationship, scholars "have become aware that the disembodied self, thinking in 'splendid isolation,' is more the product of a cultural bias than a universal phenomenon."[54] From this perspective, Chick's shift of his focus from independency to intersubjective relationships is interpreted as part of the modern process to create an alternative worldview with others and to criticize one's own obsessive demand for independence as a narrow community's erroneous habit.

ALTERNATIVE WORLDVIEWS TO THE WHITE MALE ADULTS' DOMINANT DISCOURSE

In Faulkner's works, the white male subjects inherit the Southern planter's self-centered attitude toward the world, and tend to overlook the vicious aspects of their dominance over socially marginalized people. For the modern reader who recognizes this regional issue as a metaphor of the world's unfinished democratization, this white supremacy is a perverted perspective that disregards the world's malfunction. As John T. Matthews explains in "This Race Which Is Not One" (2004), the white community in Faulkner's fiction is blind to the segregation of its socially marginalized members because of its triviality, just as the police officers in Edgar Allan Poe's "The Purloined Letter" disregard the letter because it is placed casually on a card rack. Matthews calls this "fetishistic thinking"[55]—white male adults know the racial problem is there, but because of its casual appearance, the dominant men can disregard its existence in order to focus on the more important, fetishistic purposes. Matthews argues:

> An important element of Faulkner's writing about his world involves challenging the power of fetishistic thinking. The special power of fetish depends on the visibility of its operation. Like Poe's purloined letter, the fetish goes into open hiding. The power of southern fetishes like skin color, blood, and racial

> stereotype derives from their ability to admit their fictionality while exercising their effects. It is not so much unconscious or buried, repressed or suppressed truth that worries Faulkner, but highly developed social and psychological technologies like fetish that enable the white South to perpetuate beliefs, and hence protect advantages, in the face of their evident falsehood.[56]

Racial segregation occurs when the white male Southerners' monomaniac preservation of power in their society is too intense to find any abnormity in their violence. This modern fetishism with power and its eternity appears in the system of white male dominance that regards the racial problem as a local, trivial matter. Lucas and Chick focus on their intersubjective relationship and shift their perspective from the white Southerners' subjective obsession for power to the marginalized individuals' criticism of the white naturalization of violence.

After his resolution to cooperate with Lucas, Chick meets his old friend Aleck Sander and an elderly female character, Miss Habersham. They are the unique voices that compose this minor portion of community. Aleck is Chick's African American friend of the same age, and Miss Habersham is a white spinster who was a lifelong friend of Lucas's wife, Molly. These characters, who are willing to act on Lucas's behalf, create "an alternative community consisting of blacks, women and children," or "a community of minorities, marginalized on the threshold of the dominant community of adult, white men."[57] This socially marginalized group creates their common ground with antipathy toward white male, fetishistic dominance.

In *Faulkner's Rhetoric of Loss* (1983), Gail L. Mortimer rightfully argues that Faulkner's identity as the descendant of a decent Southern family is important because Faulkner's consciousness of the Southern aristocrat provides a "perceptual reference point from which we can understand why things in Faulkner's world are viewed as they are."[58] In this perspective, socially marginalized people are frequently represented by African Americans and females. Mortimer argues that Faulkner's white male characters insist on their independence from African Americans and females in spite of the fact that they inevitably depend on these social members:

> In Faulkner's stories the black person and the female evoke strikingly similar emotions in characters to whom they represent all that is "other" than the self. Defending against a closeness (= attraction = dependence) to them that feels personally threatening, the white male secures his sense of self only by repeatedly denying his ambivalence toward them, by asserting his autonomy.[59]

Faulkner's white male characters insist on their autonomy to defend themselves from their inevitable relationships with African Americans and women. Southern society is dichotomized into white male adults on one hand and women and African Americans on the other in order to defend the white males' independent minds from the unnecessary psychological wounds in their relationships with the others.

In contrast to the usual pattern of the white male characters in Faulkner's works, in *Intruder in the Dust* Chick contacts those who belong to minor social layers in order to understand that social diversity also produces interpersonal, equal relationships. Faulkner's depiction of the temporal alliances between a young white male, an African American boy, and an old spinster suggests the decentralized formation of egalitarian relationships in the American South. This incessant effort to realize equal relationships under the pressure of a dominant power's stratification of society is still necessary even today—as Hirschkop argues, "Democracy is no longer an unequivocal ideal, it is also a historical fact, not just the prize but the battleground on which social struggles take place."[60] Even if people live under a democratic government, their personal efforts to have equal relationships always struggle in a society that fundamentally creates social strata and heterogeneity.

Chick's social combinations with socially marginalized people are personal and uniquely identified. Chick and Aleck were friends from their childhood. Aleck casts off his passive African American attitude and chooses to continue his cooperative relationship with Chick, believing in the possibility of attaining social justice in Lucas's case. When Chick comes to Aleck's house, Aleck and his family are "sitting quietly in the dark behind the closed doors and shuttered windows" to avoid the white mob's fury (ID 83–84). Chick and Aleck are at an age where they are conscious of their social differences. They sense their relationship cannot be labeled as childhood friends anymore, yet Chick calls Aleck his companion when he confronts his fear of exhuming the body of Vinson Gowrie at night. As Li argues, "Chick's conscious disbelief that he will actually have to fulfill Lucas's request is ironically juxtaposed to his inability to envision himself acting alone."[61] Ridden with the idea of independence, Chick recognizes that his desire for a companion comes from his shameful powerlessness to do it alone. He does not expect Aleck to come with him, thinking "if he were Aleck Sander he wouldn't come out of the house to anybody's whistle tonight" (ID 84). But when Chick whistles their familiar signal, Aleck quickly comes out of the house. Young Aleck refuses to accept the frustration of retreat and chooses to cooperate with Chick and give priority to their intersubjective relationship. Aleck is not looking at Chick, but "looking past and over his head toward the Square" (ID 85), suggesting that he is looking at the community in which the white male adults maintain power. When Chick tells Aleck that Lucas wants him to dig up Vinson's body to prove his innocence, Aleck reacts in the same way Chick did with Lucas: "Me? Go out there and dig that dead white man up?" (ID 85). Apparently, Aleck distrusts his ability to resist the white male dominated community, but his intersubjective conscious with Chick indicates his will to take part in this matter.

Aleck's evaluation of his social combination with Chick is expressed in his rejection of the monetary reward. Chick suggests that Lucas may pay for Aleck's work, but Aleck simply shows his will to help Lucas without any expectation of money: "I aint rich, . . . I dont need money" (ID 85). His words imply that his social position has no hope to improve if the American South continues its racist regime. A young African American's contemporary desire for social equality is expressed in his disregard for potential income. His priority has already shifted to his personal relationships with Chick and Lucas; his resolution is illustrated by his contradictory utterance of his desire for money. As Richard Godden argues in *William Faulkner: An Economy of Complex Words* (2007), every speaker's voice follows some social assumptions and purposes, or metanarratives, of its users, but such a voice will include traces of other voices within it, and an individual's voice "expresses a divided or, in some circumstances, a contradictory intent." Godden quotes Bakhtin to say that, because of this "divided nature of the socially indexical sign," the nature of words and voices are always shifting and changing according to their social communication.[62] In Aleck's case, his demand for the realization of interpersonal equality in the community is expressed in one contradictory expression—he is not rich, but he doesn't need money.

Miss Habersham is a woman who has her own vocabulary and material to resist the white male dominance. This white, old spinster harshly criticizes the white men who cannot change their rigid social structure. She is a "wealthy descendant" of "one of Jefferson's founding fathers," and "something of an aristocrat."[63] Miss Habersham and Lucas's dead wife, Molly, were "born in the same week and both suckled at Molly's mother's breast and grown up together almost inextricably like sisters, like twins" (ID 86). Because of her strong compassion for Lucas's family, she voluntarily comes up to Chick and Aleck to ask how Lucas's case is going. She feels Gavin has been dealing with law and order too long to "focus on the logical surroundings of an event."[64] Miss Habersham is frustrated with the authorities who do not want to listen to Lucas: "Of course. Naturally he [Lucas] wouldn't tell your uncle. He's a Negro and your uncle's a man" (ID 88). She suggests that the white male elite, who have focused on the control of their power, have lost the ability to shed new light on their society's unknown areas.

Chick perceives that Miss Habersham's words are the "simple truth" that is "intact" in "the chronicle of man" (ID 88). She is blunt about the world's darker aspects. Moreover, she materially offers her truck to bring them to the graveyard. As Gavin's personal confession shows, ordinary Southerners at that time considered the automobile to be the "sex symbol" of white

American males (ID 233). Speaking the subversive truth openly and driving a male sex symbol, Miss Habersham goes far beyond her ordinary gender role to cooperate with Chick and Aleck. Each of these socially marginalized characters demonstrates what Bakhtin calls "a unique speech-thing" (DN 299) that can resist the white, male-oriented community's norm. Chick learns alternative views on his community from his connections with these socially marginalized characters.

Chick's social combination with the African American and the white spinster provides the central plot of this novel. To resist the dominant opinion that creates a disadvantageous condition for Lucas, these three socially marginalized characters practically find evidence that someone else is the murderer. Chick, Aleck, and Miss Habersham visit the graveyard at midnight to dig up Vinson's body, but they find someone has already dug up Vinson's grave and refilled it with Montgomery's dead body. This result is reported to Gavin and Sheriff Hampton to get them to renew their investigation of these murders. Gavin and Hampton suspect Crawford Gowrie, Vinson's elder brother, as the killer of his lumber-stealing partners. After killing his brother, it appears that Crawford murdered the other member of their group, Montgomery.

The investigation proves the murderer's cruelty. Crawford buried Montgomery's body in his brother's tomb first and then dropped Vinson's body in quicksand, where the body would automatically sink and be buried. The clumsy disposal of the bodies suggests impulsive violence rather than premeditated. Losing all respect for the unity of family, Crawford did not treat his brother any differently from his criminal partner. If it is true that, as some scholars say, respect for the sanctity of blood kinship is the very basis of the South's moral code,[65] then Crawford's indifference to the bodies suggests his murderous drive does not have any moral constraints.

Before the identification of the murderer, Chick believes the serial killer murdered two men simply for the sake of his violent "compulsion" (ID 114), inventing "his motive and reason afterward" so that he can still "stand up among man as a rational creature" (ID 115). Chick then juxtaposes the serial killer with "Vinson Gowrie's kin and neighbors" who "have to kill Lucas or someone or anyone" so that they can "lie down and breathe quiet" (ID 115). The white mob, which includes members of the community from various classes, gathers to share this impulsive expression of the poor whites' moral collapse. Chick comes to understand that serial killing and lynching emerge from the same irrational urge.

Chick finds his new cooperative conditions with Aleck and Miss Habersham are effective to find and resist the South's moral decay. When Chick cooperates with Aleck and Miss Habersham to investigate Vinson's grave at night, Aleck's keen eyes witnessed the murderer bringing Vinson's body to its grave. In the darkness, Aleck watches the murderer riding a mule and toting the body "on the saddle in front of him" (ID 98). And Miss Habersham prevents the mob's lynching by sitting on the staircase of the jail, believing that "a white lady" can stop "men with guns" (ID 115). The intersubjective connections between these three characters help them solve the murder case and keep the lynch mob from Lucas.

Two murder cases reveal that the community's hatred for Lucas is a fundamental malfunction of the imperialistic dominance of the white community. Compared to the community's deeply rooted, violent urge that dominates both the homicide and the lynching, the social combinations of these three marginalized characters seem to be fragile and temporal. Chick, however, is aware of the effectivity of their temporal, personal cooperation; he recognizes that people's will to take part in the world, to share the same sensations with others, are common discursive phenomena to solve the problems and to make their situation better. The unique achievement of *Intruder in the Dust* is Chick's positive recognition of his temporal, intersubjective relationships with others. While he explores the murder case of Vinson and Montgomery, Chick finds a way to appreciate the value of his marginalized relationships he has ignored.

SOCIAL COMBINATIONS TO LEARN THE WORLD IN DIFFERENT WAYS

Chick feels his mind's instability from the beginning to the end of this novel. In his investigation of the murder case, Chick finds his social combinations can resist against the community's violent urge, but at the same time he wants to stick to his unification with the white male voice, demanding to be a part of "one people one heart one land" (ID 205) of Yoknapatawpha County. Chick's uneasy position reflects his community's confused condition. The American South in the 1940s is in the middle of a change in moral codes. People cannot predict which direction the world will go. The early twentieth century had "African Americans heading to the urban North and white country people moving into town, both pursuing a better life," and World War II saw "a massive movement of people within and out of Mississippi."[66] In pursuit of a better life, the Southern white community's uniformity and racial exclusivity were seriously challenged. Behind Crawford's selfish murders, some critics see an increase in the South's "Northernization"[67] or the "inhuman"[68] conditions of modern life. These interpretations suggest that modern capitalism in the middle of the twentieth century increased chaos in the American South and did not give a new moral foundation. Desire for money demolished the Southern white's unity and moral superiority.

This moral confusion of people's private lives relates to the political struggle between the United States and the USSR. As Matthews argues in *William*

Faulkner: Seeing Through the South (2009), in the international condition after World War II, "the emerging struggle for global domination between the USSR and the US"[69] generated deep terror among nations and caused the US imperialism to resist the threat of communism. The chaotic conditions inherent in such a huge political struggle are reflected in Chick's mind—he explores the future of his personal connections under the huge pressure of the imperial power of the white male community. As all community members are neither safe nor steady in their positions, Chick seriously considers on what basis people can judge the appropriation of their temporal deeds. He wonders about the goal behind their actions: "We dont need to worry about stopping now. It seems to me what we have to worry about now is where we're going and how" (ID 206). Each individual's personal pursuit of prosperity causes uneasiness in people's social positions and their dialogic relationship. As Hirschkop says, "The logic of democratic intersubjectivity includes an element of *risk* for all: there is no telling where things will go."[70] The locals pursue their best judgments in their own ways, but their solutions lose any confidence in their righteousness.

Chick's recognition of this uneasy position in his community makes a sharp contrast with Gavin's reductive logic and self-confidence. Gavin appears in this novel as a Chick's mentor, but he betrays his limitations in his racist attitude toward African Americans. Gavin calls Lucas "Sambo" (ID 151), applying an anachronistic image of an African American who can endure his unfair condition while waiting to be free and equal in the future. In his attitude toward Lucas, Gavin acts as a conscientious white man who can provide patriarchal protection to African Americans. As Williamson argues, since the end of Reconstruction in the 1870s, Southern white elites are ready to agree that they are "God's chosen people to take care of Southern blacks"; Gavin inherits this tradition and so tries to pin down Lucas as "the child-like Sambo image."[71] As John W. Blassingame argues, "the Sambo stereotype" is rooted in the most antebellum whites' firm belief that African Americans are "ignoble savages" who are "innately barbaric, imitative, passive, cheerful, childish, lazy, cowardly, superstitious, polygamous, submissive, immoral, and stupid."[72] Gavin uses this historical context to maintain his superior position over Lucas. As Dimitri says, "There is apparently no question in Stevens's mind that middle- and upper-class whites are best suited to educate the black—'Sambo,' as he puts it—on the proper way to conduct the struggle for equality."[73] Even if his final purpose is the demolition of racial discrimination in the American South, his stereotypical presupposition about African Americans requires Gavin to maintain his ascendancy over Lucas. He feels that Lucas threatens his community because, after being saved from the lynching, Lucas becomes a powerful figure whose very existence calls attention to white racism. Gavin makes a sarcastic remark: "Lucas Beauchamp once the slave of any white man within range of whose notice he happened to come, now tyrant over the whole county's white conscience" (ID 195). He does not want to let go of his dominant position over other races. Gavin's reaction reveals the white Southerners' typical antipathy toward the African Americans' acquisition of social status and strength.

As Gavin presumes white people's moral superiority over African Americans, he takes it for granted that the moralistic white people regret their wrong judgment when they find they have "been wrong" (ID 194) and feel a conscious burden to Lucas. Gavin assumes this burden and claims Lucas "will receive installments on" this wrong arrest "for the rest of his life in this country whether he wants them or not" (ID 194). In Lucas's case, the white Southerners' moral superiority is reversely expressed as their moralistic debt to African Americans. Lucas becomes the symbol—the creditor—of the mistake of the American South and the white people of the community become the debtors. Following the idea of "the white man's burden"[74] that circulated in Southern intellectuals' minds since the nineteenth century, Gavin insists the white people's moral superiority should be proved by their debt to African Americans.

In contrast to Gavin's logic of white supremacy, Chick finds that Gowrie's serial murder and the white people's threat of lynching are motivated by the same moral decay of the white community. His detective process reveals that the community's prejudice against African Americans supports Gowrie's plan to make Lucas a suspect. The night the murder occurs, Lucas is out walking and witnesses Crawford Gowrie and his younger brother Vinson sneaking lumber from their uncle's property. They are secretly selling the lumber to their co-conspirator, Jake Montgomery. When Crawford realizes Lucas has witnessed his secret trade, he challenges Lucas to shoot a stump "from fifteen feet away" (ID 221) to prove that his white grandfather's antique pistol still works. After Lucas shoots the stump, Crawford shoots Vinson in front of Lucas. As a result, Lucas is found standing beside Vinson's body with a discharged pistol. Since no one believes Crawford could kill his own brother, Lucas is arrested for the murder. Lucas, knowing that blacks cannot openly blame the moral corruption of the whites, asks for Chick's assistance.

Crawford frames Lucas for the murder of Vinson to satisfy white people's social expectations. Vinson's family members and the white mob select Lucas as the victim of their lynching to satisfy white society's desire for revenge. In the process of their investigation, Chick begins to understand how the community's "biological and geographical psychology" (ID 223) prevents the proper investigation of this murder case and identifies the wrong man. The white members of the community assume the African American can shoot a man "*in the back*" and then "*sleep like a baby*" (ID 57). Crawford's serial murder starts by capitalizing upon this common presupposition of white men.

When Vinson's fratricide is revealed publicly, the people who gather to witness the lynching of Lucas cannot bear to admit their own psychological participation in a poor white family's moral degradation. When Crawford is finally identified as the murderer, the white mob recoils to their homes. Chick feels the mob returned to their homes because they realized "they were wrong" (ID 192). Chick concludes: "They were not running from Crawford Gowrie or Lucas Beauchamp either. They were running from themselves. They ran home to hide their heads under the bedclothes from their own shame" (ID 198). The white mob gathered only to find that their cruel desire for the violent oppression of African Americans had been used to support a disgraceful murder. They escaped from the recognition that their prejudices had complicity in Crawford's serial murders.

After the solution of Vinson's murder case, Chick talks with Gavin to try and understand the meaning of his own personal connections with socially marginalized people. In this dialogue, Chick is aware of his strong tendency to imagine his personal connections as his own new belonging. However, after witnessing the white mob's moral corruption, Chick does not want to belong to this alternative community because such exclusive identification gives him one monocular vision on the fact again. He even criticizes his own connections with Aleck Sander, Miss Habersham, Uncle Gavin, and Sheriff Hampton when he imagines that "they should be perfect because they were his and he was theirs" (ID 205). He knows this self-justification of himself and his group has a tendency to conceive the "furious intolerance of any one single jot or tittle less than absolute perfection" (ID 205). Chick feels his tendency to justify his own personal connections with a particular social group is "worse" than just "wrong" because in this sense of belonging he perceives himself "righteous" (ID 205). For Chick, life is now not the continuation of his own narcissistic sameness with others, but a series of critical examinations of his own identity and belonging.

Gavin, whose mind is full of pride for the white elite but conscious enough of his own limitations to find the advantage of his nephew's self-criticism, encourages Chick to continue his communication with the others and make new and better formations of social combinations. "Just dont stop" (ID 206), Gavin says to suggest Chick's communications with the others are a process of struggling to form an alternative social structure.

In his dialogical relationships with others, Chick's idea still has variability in meaning because his idea is connected with the other community members' words. Chick cannot guess the future by his own words; he reveals that his words only come from his intelligence's combination with others. The future is not what he has already conceived in his mind, but something that will be altered by his communications with others. Faltering, "—only there was something else too—I was trying . . . ." (ID 201), Chick believes the future comes from "something else"—not from his own self-centered presupposition, but from his intersubjective relationships with others in the community. These relationships imply that Chick uses Gavin's words to support his own reasoning. He cannot express his own logic without the help of Gavin's words. He feels that he has "no clothes nor expression nor talking either smoke-screened anything from his uncle's bright grave eyes" (ID 201) because his logic comes from his sharing words with Gavin.

In today's context, Chick's worldview can be understood as the local people's effort to gain better judgment on their problems by sharing the same sensation with the world. Because he cannot find any moral confidences in traditional society, Chick uses his personal solidarities and shared sensations with the world to support his own moral judgment. He explains this sense of solidarity with the world using the metaphor of eating. Borrowing Gavin's words, Chick explains that his sense of intersubjectivity goes along with his demand to share some common living sensations with others, such as eating. "Remembering how his uncle . . . had said that man didn't necessarily eat his way through the world but by the act of eating and maybe only by that did he actually enter the world, get himself into the world: not through it but into it," Chick imagines that he feels the "vast teeming anonymous solidarity of the world" using the metaphor of "the act of eating" (ID 202–3). This logic is natural for Chick. Because his experience of solidarity starts when he shares food with Lucas, Chick recognizes that the primitive experience can be a proper metaphor for him to connect his moral sense with the world. Chick understands his own basic sense of solidarity with the world can be preserved by sharing a personal and primitive sense—eating—with others. This seems to be rhetorical logic in the 1940s, but for today's international readers—who commonly share their various everyday aspects and words on a digital network—Chick's sense of sharing living sensations with others is applicable to the local individual's solidarity with the globalized world.

What Chick chooses is not to identify himself with the monolithic unity of the white male community but to keep his personal relationships with socially marginalized members, sharing some living sensations that are common in the world. In his imagination of the world's solidarity through his own primitive sense, Chick predicts that local people will be connected globally. If we place *Intruder in the Dust* in Faulkner's contemporary world of the 1940s, as Joe Karaganis argues, readers can see Chick's sense of social connections among individuals as an idea that will be developed as an imaginarily connected resistance against the Cold War logic. Later on, Faulkner crystalizes this sense of solidarity in soldiers' resistance against the war's imperial impulse in *A Fable* (1954):

> The political sequence outlined in *Intruder* [*in the Dust*], from the individual to the minority within a race, to the confederation of such minorities, to, finally, their dominance of the nation is the defense of individual conscience in increasingly large *combinations*. After *Intruder in the Dust*, this is the idea that Faulkner increasingly brought to bear against the cold war logics of authoritatrianism and militarism, both at home and abroad. It is the territory, in particular, of *A Fable*, which recounts the almost viral spread of the idea of peace among troops on the Western Front, enabling each soldier to lay down his weapon as if in perfect association with soldiers up and down the line and across the battlefield.[75]

What Karaganis claims here is that Faulkner's imagination enables the expanded resonance of conscience in individuals' personal "*combinations*," to make it resist both the nation's and the world's imperial impulse. This resonance in people's social combinations is what Faulkner illustrates in his later works, and what we are now witnessing in the people's interactive connections on the digital network.

Faulkner's imagination directly connects the personal combinations of local people with the solidarity of the rest of the world. This logic is metaphysical in the 1940s, and it has been interpreted as an idealistic expression that represents the particular generation's budding civil rights movement. For example, Charles S. Aiken argues that Chick's "generation bears the brunt of desegregation and equal rights for blacks more directly than any other."[76] Aiken interprets Chick as a representative of his generation's sympathetic sentiment for civil rights movement; however, for today's ESL students, Chick's demand for resonance between the local and the global is even realistic because ESL students can read Chick's social combination as the beginning of a local boy's desire to expand his egalitarian connections beyond the world's deeply stratified condition through the people's network. Chick can be understood as their contemporary. Chick's expansion of his own sense of community is similar to the user's sense of shared sensations in the digital space, where—as Brabazon argues, quoting Raymond Williams's maxim—"the process of communication is in fact the process of community."[77] From this point of view, the community always shifts its forms according to the development of its personal communications with others. This is another type of fashion to create their connections, and this way seems more proper to develop the equal relationships among individuals. Today's readers who have cultivated this sense of literacy in the digital network can interpret Chick as a person who forms personal relationships based on the shared sense of words and living sensations.

In the latter part of this novel, Chick feels that his society expects the socially marginalized people's new connections to improve the local community. When he walked "the same street almost exactly as they had walked it,"

Chick felt "it was not empty at all now, deserted and empty of movement certainly running as vacant of life from street lamp to street lamp as a dead street through an abandoned city but not really abandoned not really withdrawn but only making way for them who could do it better, only making way for them who could do it right" (ID 204). Chick imagines that the world's problems are not solved by one particular social group's dogmatic moral, but by some who have multiple, equal connections with people to have various points of view. He walked in the empty street feeling his own personal connections with multiple others as if he were our contemporary who keeps his digital connections alive while he walks in the dark street. Chick feels that such multiple, equal connections that are suggested in his communications with the socially marginalized people can be appropriate to search for the better solution.

Intruder in the Dust also demonstrates how people can easily lose their moral connections with the world when they do not have the will to attain equal relationships in their activities. Chick senses the world loses its moral sense when it does not want to maintain interpersonal equality. After the mob disappears, the sheriff decides to use Lucas as a human bait to arrest Crawford—dangerous and inhuman—but no other solution can be found. Watching the sheriff lead Lucas, Chick imagines that in the future he will remember how he felt the town's immoral condition was isolated from the world:

> And he would remember: himself and his uncle standing beside the sheriff's car in the alley beside the jail watching Lucas and the sheriff emerge from the jail's side door and cross the dark yard toward them. It was quite dark in fact since the street light at the corner didn't reach this far nor any sound either; only a little after ten oclock and on Monday night too yet the sky's dark bowl cupped as though in a vacuum like the old bride's bouquet under its glass bell the town, the Square which was more than dead: abandoned. (ID 206)

The community is disconnected—or "abandoned"—from a global morality when the community members conduct their actions without democratic principle. What Chick perceives here is the fact that, once people lost their sense of personal connections based on egalitarian relationships, there was no principle that could guide them because they lost their confidence in the fairness of their actions.

Chick evaluates the community members' efforts to keep equal relationships among individuals, even if it is difficult to be considered as the truly proper solution to a problem. Since the local community is the place where personal judgment is created based on a few people's interpersonal equality, its judgment is subject to the outer world's criticism. The final judgment on the murder case in Chick's community is conducted not based on official law but on the personal communication between the sheriff and the murderer.

After Crawford is arrested, Sheriff Hampton gives Crawford a Lugar pistol with one bullet in it, suggesting that Crawford should kill himself before the shameful trial of fratricide. Hampton believes that the white male Southerners' sense of honor can be the common ground on which he can establish proper justice for the murderer. For Hampton and Crawford, the gun is a tool with which to communicate the traditional sense of honor without words. Kenneth S. Greenberg explains that during the Civil War, some Southern men considered suicide by pistol an honorable death because it showed their mastery and control of their lives rather than dependence and submission.[78] The sheriff's arrangement succeeds. As the pistol activates the white gentleman's sense of honor in Crawford's mind, it is reported that he committed suicide in the sheriff's custody. Their mutual understanding of the Southerner's sense of honor works as a local ethic by which to solve the problem. Hampton's activation of a historical tradition reveals that the Southern sense of honor coexists with brutal violence in Crawford's mind. Making an ancient tool modern, the sheriff suggests that this historical legacy can create equal communication among individuals. This local, outdated solution cannot escape the criticism of the rest of the world. Chick thinks "strangers" may frown on this arrangement and ask "what sort of jail and sheriff" exist in Yoknapatawpha County (ID 231). However, at the least, here the past is evaluated when it becomes not an absolute rule to be obeyed by the people but a legacy to be used to give mutually agreed-upon moral judgment.

Intruder in the Dust suggests that individuals can have social combinations to relocate their positions in the world. If racial segregation is the result of the traditional social strata of the American South, it is effective to reconfigure individuals' mutual connections through their own temporal, equal relationships. Individuals share their living sensations in their daily communications, and at the same time they perceive their mutual desire to make their relationships better. These two elements of people's personal connections are indicated in Chick's intersubjective communications with the socially marginalized members.

After his experiences with social combinations beyond the boundaries of race, gender, and class, Chick thinks Gavin's paternal logic cannot suppress the individual's demand for connections beyond social borders. After the resolution of the Crawford case, Gavin orders Lucas to visit Miss Habersham with flowers to express gratitude for her help. Gavin's main purpose is to give an uncomfortable task to the arrogant, elderly man. As Gavin expects, Lucas is embarrassed: "I aint much of a visiting man" (ID 235). Gavin succeeds in imposing a polite form of harassment on Lucas, but, unexpectedly, his imposition inspires Chick to accompany Lucas. Worrying that some whites may still hide in the bush to take revenge, Chick goes with Lucas as his white companion. Near the end of the novel, Chick finds he can still cooperate with

Lucas. They become two visiting men who make a new path to avoid the community's racial malice.

Chick focuses on society's intersubjective aspects in order to discover that socially stratified individuals can be connected through common elements in their daily life. Near the end of this story, Chick perceived a funny association between Lucas and Miss Habersham. Lucas took "the same hard chair beside the door which nobody else but Miss Habersham had ever chosen" and used "a leather snap-purse patina-ed like old silver and almost as big as Miss Habersham's handbag" (ID 235, 238). In spite of the apparent differences in race, social position and gender, these two elderly Southerners share the same generation's habit and fashion. A historical legacy is reconsidered as a material that can connect members of different social groups. Chick's worldview releases characters from the South's racial chains and catches them again in a more interrelated atmosphere.

The characters' social combination implies a future connection. In the final conversation between Lucas and Chick, Lucas states that his familiarity with Chick will continue in the future. Lucas humorously asks Chick, "You aint fell in no more creeks lately, have you?" Chick responds, "I'm saving that until you get some more ice on yours." Lucas warmly remarks: "You'll be welcome without waiting for a freeze" (ID 235). This conversation can be a metaphor: Lucas hopes Chick will not freeze his identity as a member of the white male community, but will keep his relationships with the socially marginalized people in order to maintain their society's flexibility. Lucas's hope suggests that if they actually have temporal equal relationships several times, they can still believe their social positions are free from rigid stratifications.

Faulkner offers Chick's intersubjective relationship with Lucas as a Southerner's ability to have a personal relationship with someone outside the norm. Historically, in spite of the budding civil rights movement and its demand for progress, laws of citizenship, employment, and educational opportunities betray the American South's dismal record and sluggish backwardness.[79] As Moreland argues in "Contextualizing Faulkner's *Intruder in the Dust*" (1997), three overlapping changes—the end of the age of Europe, America's ascent as an absolute world power, and third-world decolonization—make the Southern racial segregation of the 1940s a noteworthy matter for the world.[80] Faulkner describes Chick's hateful reaction to lynching not only as a young Southerner's autonomous resistance against his own community's desire to dominate African Americans, but also as his personal discovery of the opportunity to redefine his position in his personal relationship with Lucas. The white male adults in the American South have not created a culture of dialogical relationships with African Americans; however, Lucas and Chick's social combination demonstrates their ability to have a dialogical space that resists the dominant community's annihilative attitudes.

With *Intruder in the Dust*, Faulkner's dominant power as an author seems to decline. Once the equal relationships among individuals are discovered as the goal of his literary work, Faulkner cannot keep his privileged position as an author. Faulkner's limitations, however, are effectively used to depict the white protagonist's demand for social combinations to gain a new perspective about their community. In his later career, by narrating that moral judgment comes from the author's and characters' historically and geographically limited conditions, Faulkner reveals that an individual's sole effort is unable to solve the American South's racial problem.[81] Faulkner—a conservative, white male Southerner—could not agree with the sudden abolition of the discriminations between races. His conservative attitude toward the racial problem is obvious in his later career. In his notorious essay, "Letter to a Northern Editor" (1956), Faulkner requests people to "go slow now"[82] on the racial issue, suggesting he does not believe the complete assimilation of both races is an ideal goal. His conservative stance expects the African American leaders' adaptable approach to his local area's racial issue. In "A Letter to the Leaders in the Negro Race" (1956), Faulkner demands that "not the individual Negro to abandon or lower one jot his hope and will for equality, but his leaders and organizations to be always flexible and adaptable to circumstance and locality in their methods of gaining it."[83] Faulkner's personal opinion demands that the African Americans' autonomous activity to improve their situation should be "flexible and adaptable" for the dominant people, and reveals his white-male-oriented, conservative attitude toward African Americans.

Faulkner objectively examines his own position and depicts his ego's historical limitation using his characters in *Intruder in the Dust*. Faulkner's Southern white male characters cannot solve the racial issue by themselves when they believe that they are the only group who can solve it. Like Gavin Stevens in *Intruder in the Dust*, the white male characters assume they are the problem-solvers and the marginalized characters are the troublemakers. The white male elites reveal their own imperial impulse even when they want to solve the racial problem. Faulkner, who personally wants to solve his contemporary race issue retaining his supreme position as a decent white in the community, uses his talent to create a dialogical space in which heterogeneous characters make social combinations to solve the problem in *Intruder in the Dust*. Faulkner's honest pen does not hesitate to express that Chick's mind still retains the white elites' arrogant aspect, but the young protagonist's social combinations with the socially marginalized members allow him to have other viewpoints. Lucas and the other socially marginalized characters—an accused African American, teenage boys, and an isolated spinster—are able to have their own temporal relationships that cannot be completely accepted by the author's conservative mind.

What Chick finds in this story is not the community's traditionally arranged positions, but an imagined space where each individual's broad range of social functions can be connected to compose dialogic relationships among various socially stratified members. Comprehending his community as an incomplete process of formation, Chick finds that Lucas is a living person who can still change his attitude toward others. At the end of the novel, Lucas comes to Gavin's office to offer payment for his case. Gavin, wanting to keep his superior position, refuses Lucas's payment. He suggests that an exchange of money is not the proper expression for their relationship. Gavin intentionally refuses Lucas's payment in order to burden Lucas with a psychological debt toward a member of the white elite. Additionally, Gavin requires just two dollars to compensate for the fountain pen he broke while he was writing down Lucas's information for Hampton. He requires such a small amount of money to confirm his patriarchal position over Lucas. Lucas obeys Gavin's request, but he tries to recompose their relationship as a mutual understanding among individuals. Quoting a music box's lyrics, "to try to learn you different" (ID 239), Lucas defines himself as a person who is willing to learn different procedures from others. Following Gavin's insistence, "This is business" (ID 240), Lucas counts out the correct coins to pay his two dollars. It is a satisfying conclusion for Gavin, but not for Lucas. What Lucas desires in his conversation with Gavin is to have a common ground on which they can deal as equals for a while. After his payment, Lucas requests a "receipt" of this business deal (ID 241). Earlier in the novel, Lucas shows off his blood relationship with a respected white family to maintain his aloof attitude toward the other community members, but now he negotiates with Gavin to reaffirm his position as an individual who can change his stance according to the flow of circumstances. This last word of Lucas suggests the American South's progress can be possible when participants of democratic dialogue add another viewpoint to the world.

CONCLUSION

Reading *Intruder in the Dust* as an introductory work for Faulkner's novels is beneficial to ESL readers whose digital network has expanded to the world. Faulkner uses Chick's intersubjective relationships with others to suggest that the moral foundation of the community is retained not by the heroic subjects' obsessive worldviews, but by the socially stratified individual's will to have equal relationships, even if it is fragile and temporal. Being aware of the international discussions on racial problems, Faulkner imagines that the local people's social combinations resonate with the world's discursive struggle to have egalitarian relationships with others. Analyzing Faulkner's text as a metaphor of today's digital network and communications, ESL students can read Chick's attempts as a process to have personally democratic relationships while retaining each individual's social mobility and moral uncertainty. Social combination causes moral uneasiness among individuals, but this novel suggests that people's intersubjective egalitarianism inspires their demand for suitable conditions in their mutual relationships. Faulkner offers this novel to show that personally democratic attitude in the American South is part of a global process to create social formations and the people's new self-fashioning to realize equal relations among individuals.

Intruder in the Dust reflects Faulkner's matured idea on the race problem in the 1940s. Faulkner depicts that the Southerners confront the large-scale mobility of the population and they needed to reconfigure their sense of communities. Crawford's serial murders reveal the desires of modern Southern white men to dominate others and offend morality. The white mob's sadistic desire for lynching and their recoiling from the truth proves the falsehood of the white community's moral superiority and unity. The Southerners' traditional moral codes do not provide the wisdom to govern their behavior. Their aloof self-fashioning cannot have proper communications with the others. In contrast, the successful results of the protagonist's dialogic relationships with Lucas, Aleck Sander, and Miss Habersham show that personal communications beyond the boundaries of race, gender, and class can improve judgment on a deeply stratified society's problems.

Chick's social combination with Lucas shows the start of a young Southerner's process to create a dialogical space in his community. He steps out of the white-male-oriented narrow community and begins to accept other voices in order to see the lynch problem in the new light. The shift of Chick's focus from his imagined independence to his contemporary relationships with others makes him discover a more critical view of the Southern white male dominance. Lucas's words remind Chick of Old Ephraim, who suggests that even if the South is deeply stratified according to race, gender, and class under white male dominance, it still retains the socially marginalized people's personal connections and dialogic spaces that can effectively work to gain what they want. The socially marginalized people are represented by the old African American Lucas, the white spinster Miss Habersham, and sixteen-year-old Chick and Aleck Sander. Aleck Sander and Miss Habersham make their social combinations with Chick to demonstrate their own critical attitudes against white male dominance. Aleck Sander shows his resistance against white dominance by cooperating with Chick without any economic reward, and Miss Habersham criticizes the white males' self-centered worldview in her words and shows her power by sitting on the doorstep of Lucas's jail to prevent the mob's entrance. Their demands for equality are explored in their ability to have and keep their personal relationships.

The main focus of the latter part of *Intruder in the Dust* is the young protagonist's struggle to find a way to recognize the local members' ethical validity that can be shared with the world. What Chick desires is not a sense of identification with a particular dominant group, but an egalitarian relationship with other individuals that will endorse his sense of solidarity with the world. Chick finds this connection avoids the white community's unfair verdict and evaluates the local people's awkward efforts to have equal relationships. For example, the sheriff gives his final judgment on the murder case based on his personal relationship with the murderer, Crawford Gowrie. The sheriff attempts to reconnect Crawford's pride with the traditional sense of honor in the old South. For the sheriff and Crawford, this judgment is based on the common ground they share. Chick discovers individual connections share their moralistic judgment with the world when they aim for democratic relationships among them, even if they sometimes appear quite local and antique.

The final judgment on the murder case is conducted not based on official law but on personal efforts to have a mutual understanding between individuals. In *Intruder in the Dust*, this sense of mutual comprehension releases individuals from their static positions and temporarily puts them on the same ground for the better formation of their social connections. Lucas's last request for a receipt in this novel suggests his will to search for a new relationship in his temporal communication. Faulkner's depiction of characters who cultivate egalitarian relationships with others is useful for ESL students to reconsider the individual communications of contemporary people as the process to create a better society and culture.

Because of its marginalized characters' dialogic relationships, *Intruder in the Dust* is an appropriate text to teach the literature of the American South to today's ESL students. As Kreyling argues, teaching the literature of the American South is effective in class because the American South is "the richest site yet discovered in the U.S. cultural terrain for the study of and participation in the reinvention of culture."[84] At a global level, every community's culture becomes marginal, and people want to enjoy their mutual differences in their egalitarian conditions. Contemporary people—who know their world's different lifestyles and moral values because of the digital network—recognize they cannot have a complete assimilation, but they find egalitarian relationships are proper attitudes for their communications even if they are temporal. In his study of digital dialogue and community, Brabazon concludes, "digitisation has offered a range of proxies for social justice and information justice" because "all citizens deserve high-quality information, delivered on a stable platform, in good time and in an appropriate context."[85] In today's globalized condition, people need to develop their own social combination tactics to gain every community's cultural maturity and mutual egalitarianism. *Intruder in the Dust* effectively reconnects the struggle of characters in the American South with universal efforts to make better relationships among individuals.

In his later works, Faulkner depicts how marginalized individuals, who are generally considered as the weak and helpless, can build common ground between themselves to resist the dominant subjects' imperial power. In contrast, in his romantic works, W.C. Falkner's idealistic liberty is realized in the white males' patriarchal attitude toward the weak and helpless. The dialogical relationships of Faulkner's socially marginalized characters criticize the American South's monologic defects that are typically depicted in W.C. Falkner's works. Comparing Faulkner's works with W.C. Falkner's works is an effective way to teach the historical process of development from W.C. Falkner's works to Faulkner's. In the next chapter, I analyze how a teacher can help ESL students understand and argue both Fa(u)lkners' works in today's digital environment.

NOTES

1. Patrick H. Samway, S.J., *Faulkner's "Intruder in the Dust": A Critical Study of the Typescripts* (Troy, NY: Whitston, 1980), 1.
2. David Barton and Carmen Lee, *Language Online: Investigating Digital Texts and Practices* (London: Routledge, 2013), 68.
3. Sylvia Jenkins Cook, "Reading Clothes: Literary Dress in William Faulkner and Erskine Caldwell," *Southern Literary Journal* 46, no.1 (2013): 5, 10. Gray argues that "self-fashioning" is a common custom not only of Faulkner's characters but also of Southern culture. See Richard Gray, "Foreword," in *South to a New Place*, xvi.
4. For example, some companies enjoy their powerful positions in the genre of online search engines—*Google* in Euro-American countries, *Baidu* in China, *Yandex* in Russia and *Yahoo* in Japan—and online trading platforms—*eBay* and *Amazon*. See Justus Haucap and Ulrich Heimeshoff, "Google, Facebook, Amazon, eBay: Is the Internet Driving Competition or Market Monopolization?" *International Economics & Economic Policy* 11, no.1–2 (2014): 54–60.
5. John Egerton argues that the American North and the American South learned each other's worst parts to make a homogenized society. If we apply his logic to today's globalized condition, it is possible to see that our society's homogenization on the digital network increases the world's sharing of its worst parts. See John Egerton, *The Americanization of Dixie: The Southernization of America* (New York: Harper's Magazine Press, 1974), xx.
6. Kreyling, *Inventing Southern Literature*, xviii.
7. Samway, *Faulkner's "Intruder in the Dust,"* 1.
8. William Faulkner, *Faulkner at Nagano*, ed. Robert A. Jelliffe (Tokyo: Kenkyusha, 1956), 138.

9. Michael Millgate, *The Achievement of William Faulkner* (London: Constable, 1965), 46.

10. John T. Matthews, *William Faulkner: Seeing Through the South* (Chichester: Wiley-Blackwell, 2009), 226.

11. David Carpenter, "Faulkner's *Intruder in the Dust*: Apology for South's Jim-crow," review of *Intruder in the Dust*, by William Faulkner, *Daily Worker* (October 6, 1948): 13; Charles I. Glicksberg, review of *Intruder in the Dust*, by William Faulkner, *Arizona Quarterly* 5 (1949): 87; Granville Hicks, review of *Intruder in the Dust*, by William Faulkner, *Tomorrow* 8 (1949): 57–58; Alex Murphree, "Faulkner's Novel Offers Parable on Southerner's Duty to Negroes," review of *Intruder in the Dust*, by William Faulkner, *Post* (October 3, 1948): C4.

12. Irving Howe, *William Faulkner: A Critical Study* (1952; Chicago: Elephant Paperbacks, 1991), 98.

13. William Van O'Connor, *The Tangled Fire*, 137–38.

14. Winchell suggests that Vickery's book on Faulkner has "an antisouthern bias," but her arguments on Faulkner's works are quite correct and still useful; see Mark Royden Winchell, *Reinventing the South: Versions of a Literary Region* (Columbia: University of Missouri Press, 2006), 133.

15. Olga W. Vickery, *The Novels of William Faulkner: A Critical Interpretation* (1959; Baton Rouge: Louisiana State University Press, 1964), 139.

16. Cleanth Brooks, William Faulkner: The Yoknapatawpha Country (1963; Baton Rouge: Louisiana State University Press, 1990), 292. Cleanth Brooks's idea of the unitary condition of the Southern community comes from his strong sympathy with the Southern elites' agrarianism, which is expressed in the pro-Southern manifesto, *I'll Take My Stand* (1930). One of the central contributors to this famous anthology, Allen Tate, sympathetically considers the South is the place where people can keep their independent spirit with self-content in the pastoral world. Tate argues that the American South should "re-establish a private, self-contained, and essentially spiritual life." For Tate, the American South is the distinct place that can resist the mechanized, dehumanized modern world. Brooks, who respects Tate's agrarianism, considers that Faulkner is so deeply committed to a particular local area that his work suggests the aesthetically beautified, provincial value that is lost in modern, capitalized culture, like the works of William Butler Yeats and Robert Frost. See Tate, "Remarks on the Southern Religion," 175; For Brooks' provincialism and his preference for the agrarian tradition, see Winchell, *Reinventing the South*, 130–31; Winchell, *Cleanth Brooks*, 124, 311; Richard C. Moreland, *Faulkner and Modernism: Reading and Rewriting* (Madison: University of Wisconsin Press, 1990), 5; Richard Gray, *Southern Aberrations: Writers of American South and the Problems of Regionalism* (Baton Rouge: Louisiana State University Press, 200), 99.

17. Millgate, *The Achievement of William Faulkner*, 215–16.

18. Lewis Leary, *William Faulkner of Yoknapatawpha County* (New York: Thomas Y. Crowell Company, 1973), 172.

19. Myra Jehlen, *Class and Character in Faulkner's South* (New York: Columbia University Press, 1976), 126, 96.

20. Leary, *William Faulkner*, 172.

21. Eric J. Sundquist, *Faulkner: The House Divided* (Baltimore: Johns Hopkins University Press, 1983), 3, 26.

22. Ibid., 27.

23. Ibid., 20; John T. Irwin, *Doubling and Incest / Repetition and Revenge: A Speculative Reading of Faulkner* (1975; Baltimore: Johns Hopkins University Press, 1996.), 111–112.

24. Karl F. Zender, *The Crossing of The Ways: William Faulkner, the South, and the Modern World* (New Brunswick: Rutgers University Press, 1989), 121, 130.

25. As Sundquist argues, Faulkner views Joe Christmas—who believes that his body contains African American blood—as a negative persona of a white who is isolated from society. He assumes Faulkner's "career up to *Light in August* might well be considered as extended repression of the figure of the Negro, who truly remains a *figure* for Faulkner since his great 'black' characters are 'white.'" For Faulkner, Christmas is a negative symbol of the white male in the American South who is afraid of the sudden return of the repressed anxiety. See Sundquist, *Faulkner: The House Divided*, 26.

26. Hagood, *Faulkner's Imperialism,* 8; Smith and Cohn, "*Introduction*," 3; Gray, "*Foreward*," 7; Millichap, *Dixie Limited*, 4.

27. Kreyling, *Inventing Southern Literature*, 127, 130, ix.

28. Yaeger, *Dirt and Desire*, 96.

29. Jon Smith, *Finding Purple America: The South and the Future of American Cultural Studies* (Athens: University of Georgia Press, 2013), 91, 97, 105.

30. Since Frederic J. Hoffman's guidebook on Faulkner's works, *William Faulkner* (1961), many critics divide Faulkner's career into three parts. In Hoffman's idea, the first part is the "time of apprenticeship" (1924–1929), when Faulkner tried to decide if he would become a writer and what kind he should be. This period includes *Soldier's Pay* (1926) and *Mosquitoes* (1927). The second part is the "time of genius" (1929–1936), during which he produced such first-rate novels as *Flags in the Dust* (1973, originally published as *Sartoris* in 1929), *The Sound and the Fury* (1929), *As I Lay Dying* (1930), *Sanctuary* (1931), *Light in August* (1932), and *Absalom, Absalom!* (1936). Hoffman believes these six works are "the core of Faulkner's major achievement." The final period is the time of "consolidation and affirmation" (1940–1964), during which the presentation of Yoknapatawpha County is enriched and elaborated upon and Faulkner is "pushed far beyond it to attempt a statement of universal truths." This third phase of his career includes the Snopes saga—*The Hamlet*, (1940), *The Town* (1957), and *The Mansion* (1959); and the works that are specifically concerned with the Negro as a race, a group, and a moral problem—*Go Down, Moses* (1942), *Intruder in the Dust* (1948), and "an elaborate parable or allegory," *A Fable* (1954). This term ends with Faulkner's last work, *The Reivers* (1962), which contains a nostalgic, mellow, and even "sentimental quality." See Frederic J. Hoffman. *William Faulkner* (1961; New Haven: College and University Press, 1966), 18–20, 116.

31. Toni Morrison, *Beloved*, (1987; New York: Vintage, 2004), 322.

32. Morrison's recent work, *Home* (2013), also explores the new formation of individuals. The protagonists, a veteran who returned from the Korean War with PTSD and his sister, who lost her genital function because of an Anglo-American doctor's

eugenic operation, create a new personal connection. The veteran, Frank, lost his masculinity because of his traumatic experience in the war, and his sister, Cee, cannot have a baby anymore. At the end of this novel, Cee touches her elder brother's shoulder to go to their "*home*"—for both of them *home* means their new relationship. They know they should maintain their personal relationship in a different way, not as brother and sister, not because of male masculinity nor female childbirth, but as individuals who equally want to recover from victimized conditions. In *Home*, the characters' social combination is evaluated as a new beginning for the characters' fundamental social formation. If this evaluation of the personal relationship is reflected in the history of the literature of the American South from the nineteenth century to the twenty-first, today's critics will discover the literature of the American South always admires various heteroglot characters' social combinations. Moreover, it is possible to find intersubjective relationships in literary works that are not considered Southern. Louise Erdrich's *The Round House* (2012) depicts two young Native Americans' judgment in today's minor community. The young protagonist Joe and his friend Cappy shoot the man who raped and tried to kill Joe's mother. Joe recognizes himself and his friend as a combination of "Me and Cappy" and creates a new formation of social ethic and structure. Morrison and Erdrich both illustrate the possibilities and dangers of social combinations and explore the right way to form new social connections. See Toni Morrison, *Home* (New York: Vintage, 2013), 147; Louise Erdrich, *The Round House* (New York: HarperCollins, 2012), 305.

33. Hagood, *Faulkner's Imperialism* 9, 30–31, 145, 147, 31, 29.

34. Ibid., 150–51.

35. Sigmund Freud argues only the new paradigm can make people free from the old symbolic order. In *Moses and Monotheism*, Freud suggests that Moses's discovery of the new paradigm (God) can release the Jewish slaves from Egypt. To make people move from one stage to another, a new symbolic formation, or paradigm, must be found for their purpose. In Chick's case, he guides people to the egalitarian relationships in which the monolithic ideology cannot govern. See Sigmund Freud, *Moses and Monotheism*, in vol. 23 of *The Standard Edition of the Complete Psychological Works of Sigmund Freud,* trans. James Strachey (London: Vintage, 2001), 1–138.

36. William Faulkner, "On Privacy," in *Essays, Speeches and Public Letters* (New York: Modern Library, 2004), 65.

37. Hirschkop, *Mikhail Bakhtin*, 37.

38. Matthews, *William Faulkner*, 226.

39. In *From Tobacco Road to Route 66: The Southern Poor White in Fiction* (1976), Cook argues Faulkner finds the poor white as the important factor to reconsider white superiority in the South. In *American Hungers* (2008), Gavin Jones concisely summarizes Cook's argument that Faulkner "refuses to reduce poor individuals to the sum of their socioeconomic circumstances, and allows individuals an emotional and moral range that disrupts stereotype, sentimentality, and absolute class differences." Sylvia Jenkins Cook, *From Tobacco Road to Route 66: The Southern Poor White in Fiction* (Chapel Hill: University of North Carolina Press, 1976), 39–45; Gavin Jones, *American Hungers: The Problem of Poverty in U.S. Literature, 1840-1945* (Princeton: Princeton University Press, 2008), 188n27.

40. For more details on the relationship between African Americans and white elites, according to the dishes that are served in traditional plantation society, see Charles S. Aiken, *William Faulkner and the Southern Landscape* (Athens: University of Georgia Press, 2009), 196–97.

41. Richard C. Moreland, *Faulkner and Modernism: Reading and Rewriting* (Madison: University of Wisconsin Press, 1990), 9.

42. Philip M. Weinstein, *Faulkner's Subject: A Cosmos No One Owns* (Cambridge: Cambridge University Press, 1992), 86.

43. Edward L. Ayers, *The Promise of the New South: Life after Reconstruction* (Oxford: Oxford University Press, 2007), 156–58; Doyle, *Faulkner's County*, 378–79.

44. Millgate, *The Achievement of William Faulkner*, 220.

45. Donna Gerstenberger, "Meaning and Form in *Intruder in the Dust*," *College English* 23, no. 3 (1961): 224.

46. Wesley Morris and Barbara Alverson Morris, *Reading Faulkner* (Madison: University of Wisconsin Press, 1990), 235.

47. Philip M Weinstein, *What Else But Love?: The Ordeal of Race in Faulkner and Morrison* (New York: Columbia University Press, 1996), 125.

48. Weinstein, *Faulkner's Subject*, 77.

49. Gerstenberger, "Meaning and Form in *Intruder in the Dust*," 224.

50. Carl J. Dimitri, "*Go Down, Moses* and *Intruder in the Dust*: From Negative to Positive Liberty," *Faulkner Journal* 19, no. 1 (2003): 19.

51. Morris and Morris, *Reading Faulkner*, 224.

52. Stephanie Li, "*Intruder in the Dust* from Novel to Movie: The Development of Chick Mallison," *Faulkner Journal* 16, no. 1–2 (2000–2001): 112.

53. Masami Sugimori, "Signifying, Ordering, and Containing the Chaos: Whiteness, Ideology, and Language in *Intruder in the Dust*," *Faulkner Journal* 22, no. 1–2 (2006–2007): 61.

54. Hermans and Kempen, *The Dialogical Self*, 29.

55. John T. Matthews, "This Race Which Is Not One: The 'More Inextricable Compositeness' of William Faulkner's South," in *Look Away!: The U.S. South in New World Studies*, ed. Jon Smith and Deborah Cohn (Durham: Duke University Press, 2004), 204.

56. Ibid.

57. Morris and Morris, *Reading Faulkner*, 224.

58. Gail L. Mortimer, *Faulkner's Rhetoric of Loss: A Study in Perception and Meaning* (Austin: University of Texas Press, 1983), 12.

59. Ibid., 15.

60. Hirschkop, *Mikhail Bakhti*, 294.

61. Li, "*Intruder in the Dust*," 113.

62. Richard Godden, *William Faulkner: An Economy of Complex Words* (Princeton: Princeton University Press, 2007), 8.

63. Dimitri, "*Go Down, Moses* and *Intruder in the Dust*," 20.

64. Ibid.

65. Vickery, *The Novels of William Faulkner*, 138. R.H. King also argues that the American South's collective fantasy about the family determines the Southerners' "structure of feeling." See R.H. King, *A Southern Renaissance*, 27.

66. Doyle, *Faulkner's County*, 377.

67. Erik Dussere, "The Debts of History: Southern Honor, Affirmative Action, and Faulkner's *Intruder in the Dust*," *Faulkner Journal* 17, no. 1 (2001): 55.

68. David Anshen, "Faulkner's 'Curious Folk': Estrangement as Critique of Capitalism," *Mississippi Quarterly* 61, no. 3 (2008): 494.

69. Matthews, *William Faulkner*, 238.

70. Hirschkop, *Mikhail Bakhtin*, 107.

71. Williamson, *William Faulkner*, 155.

72. John W. Blassingame, *The Slave Community: Plantation Life in the Antebellum South* (New York: Oxford University Press, 1979), 227.

73. Dimitri, "*Go Down Moses* and *Intruder in the Dust*," 21.

74. Wells, *Romances*, 9.

75. Joe Karaganis, "Negotiating the National Voice in Faulkner's Late Work," in *A Gathering of Evidence: Essays on William Faulkner's "Intruder in the Dust,"* ed. Michel Gresset and Patrick Samway, S.J. (Philadelphia: Saint Joseph's University Press, 2004), 125.

76. Aiken, *William Faulkner*, 196.

77. Brabazon, "Introduction," 5.

78. See Kenneth S. Greenberg, *Honor and Slavery: Lies, Duels, Noses, Masks, Dressing as a Woman, Gifts, Strangers, Humanitarianism, Death, Slave Rebellions, the Proslavery Argument, Baseball, Hunting, and Gambling in the Old South* (Princeton: Princeton University Press, 1996), 92–95. Greenberg clarifies that most Southern men of honor do not think suicide is an appropriate mode of death, but they do have trouble with the idea of a proper Christian death that requires them to be submissive to God's will.

79. Blotner, *Faulkner*, 2: 1531–34; Allison Graham, *Framing the South: Hollywood, Television, and Race During the Civil Rights Struggle* (Baltimore: Johns Hopkins University Press, 2001), 6–11; Ian Haney López, *White By Law: The Legal Construction of Race* (New York: New York University Press, 2006), 28; Robert J. Weiss, *"We Want Jobs": A History of Affirmative Action* (New York: Garland, 1997), 7.

80. Richard C. Moreland, "Contextualizing Faulkner's *Intruder in the Dust*: Sherlock Holmes, Chick Mallison, Decolonization, and Change," *Faulkner Journal* 12, no. 2 (1997): 58.

81. Many critics argue that *Intruder in the Dust* shows the author's conservative attitude toward the racial issue in the American South. For example, Li points out that Faulkner shows his own limitations in his depiction of the American South's chaotic condition. She argues that *Intruder in the Dust* shows "the dominance of white narratives and the absence of an independent black voice." Li criticizes Faulkner's partial attitude toward the characters: by "refusing to explore the subjectivities of Lucas and Aleck Sander, Faulkner demonstrated his inability to discern the nature of a black perspective and its relationship to whiteness. Although he was the owner and sole proprietor of Yoknapatawpha County, Faulkner did not possess knowledge concerning all of his characters. In this regard, he must be considered not the creator of an entire universe, but a single, subjective participant in this one." In my argument, however,

Faulkner's inability demonstrates the intersubjective effects of literary language. See Li, "*Intruder in the Dust*," 117.

82. William Faulkner, "Letter to a Northern Editor," in *Essays, Speeches and Public Letters*, 87.

83. William Faulkner, "A Letter to the Leaders in the Negro Race," in *Essays, Speeches and Public Letters*, 108–09.

84. Kreyling, *Inventing Southern Literature*, 182.

85. Tara Brabazon, "Conclusion: White Men Rule?" in *Digital Dialogues*, 256.

Chapter 3

Social Combination in Class

How to Teach Two Fa(u)lkners' Works and Digital Literacy to ESL Students

> Concretely, one always addresses someone, and that someone does not assume a purely passive role (as the term "recipient" could lead one to infer): the interlocutor participates in the formation of the meaning of the utterance, just as the other elements—similarly social—of the context of uttering do.
>
> —Tzvetan Todorov, *Mikhail Bakhtin*[1]

Literature is one of the resources for individuals to comprehend the historical process of another region's culture. Some recent pedagogical studies discuss the ways to help ESL students explore literary texts to increase their English skills,[2] but these studies do not explain the methods to connect ESL students' dialogical activities with comprehension of these literary works' historical and cultural contexts. This chapter outlines how I contextualized and taught the literary works of two Fa(u)lkners (W.C. Falkner and Faulkner) in today's digital environment at a Japanese university. Teaching both Fa(u)lkners' works in Japanese ESL classrooms using digital systems (Moodle, WordPress, and Google Drive) helped enhance students' understanding of the Literature of the American South and its contexts in more effective ways.

W.C. Falkner's works, which have been treated as minor in the Southern literary canon, still have value when they are read as typical texts of the Southern romantic tradition in the nineteenth century. As Duclos argues, though it is readily acknowledged that W.C. Falkner's works are "among the minor popular literature of nineteenth century America and that most of them have glaring defects and weaknesses in style and content, they are nevertheless interesting examples of the second-rate literature of the period."[3] In my classes, students translated W.C. Falkner's works (*The Spanish Heroine* and

The Little Brick Church) to comprehend the nineteenth century's Southern, "second-rate" literary world. For ESL students, this romantic literary genre is worth reading and translating in order to understand what kind of elements Faulkner resisted and re-accentuated in his modern literary works. As these two Fa(u)lkners' works have contrasting views on their community and culture, comparing their works in class can encourage students to develop their ideas about the progression from one author's writing to the other's. Two Fa(u)lkners' different literary worlds inspire students to create their own historical views to explain these works' connections.

Usually, it is difficult to expect ESL students to read novels within two or three weeks and compare them in class discussions. In recent studies on teaching literature in ESL classes, teachers use a single author's short fiction or novel for their class—for example—Earnest Hemingway's short story, Chinua Achebe's *Things Fall Apart*, Kazuo Ishiguro's *Never Let Me Go*, or Raymond Carver's *Collected Stories*.[4] In these researches, ESL teachers did not connect plural authors' literary texts in class because ESL students needed to take time to read their texts closely to find the author's intentions, symbolic images, or messages. As some arguments suggest, literary work demands readers to "read carefully and validate inferences,"[5] or requests "discovering a deeper understanding of English."[6] Teachers want students to take the time to "go beyond the mere answering of questions to a discovery of what it was in literature."[7] If ESL teachers and students want to deeply explore the author's intention in class, teaching plural author's texts in a term seems to be too great a burden for students.

The literature teaching pedagogy that uses a single major author's fiction in class seems to enrich ESL students' ability to find deep meaning within the English text; however, I believe Japanese translation culture's highly developed condition requires ESL students to read more minor literature that have not been translated into Japanese. As almost all the canonical American literary works—for example, all Hemingway's works, Chinua Achebe's *Things Fall Apart*, Kazuo Ishiguro's *Never Let Me Go*, and Raymond Carver's short stories—have been translated, students can read these major works in Japanese. If ESL teachers expect to develop ESL students' reading ability using literary works as class texts, these translated works do not improve students' language and ability to understand the dense meanings of the texts in English because students are satisfied with the meanings they extracted from the translated texts. This condition is the same regarding Faulkner's works, which have been translated and well circulated in Japan. Many students read translated versions and discover the translators' deep interpretations of the texts before they tackle the original text, checking the stories, interpreting the difficult passages. It is not fair to say that students' readings of these major authors in class truly reflect their English abilities. If a teacher uses a minor work that has not been translated into Japanese along with a major American novel, the minor work can be used to challenge ESL students' real English reading ability. The teacher connects students' reading experience of minor works with the major ones to teach cultural and historical contexts. In the two classes in which I conducted my research, nineteenth-century minor works were used to focus on students' English translation training, and then twentieth-century novels were read in English using the Japanese translations for support.

Comparing the works of the same region's two authors helps inspire students' autonomous dialogical space in class. After reading both texts, students will find common cultural and historical backgrounds. Or, to compare two works of different generations' authors in the same area encourage understanding of the historical development from one author to the other. In this type of class session, students discuss their own discoveries with each other to connect the two literary works by themselves. Using two related texts, the teacher expects more dialogical communication among students, and encourages students' mutual discussions and autonomous studies.

When I used a traditional translation method in my ESL class, the classroom became dull. My class session followed a routine pattern: a student was asked to translate part of the text by standing up, reading the original text, and then orally translating it into Japanese; another student translated the next part in the same way. Translation proceeded at a snail's pace. There was almost no motivation for the students to develop their own interpretation of the text. I asked them if they had any questions on the interpretation of the text, but I rarely received any. The texts students read were generally classic works of American literature, and many students relied on Japanese translations when preparing for the lesson. In this condition, it was uncertain whether their translation activities in class honestly reflected their English abilities. My grades were uncertain, and I was tired of this style.

The ESL class activities reported in this study were motivated by my desire to change my pedagogy. Over several years, I had read nineteenth-century American literary texts that had not been translated into Japanese. For example, in my undergraduate classes, students read Elizabeth Stoddard's *The Morgesons*, *Two Men*, and her short stories; Caroline Lee Hentz's *The Planter's Northern Bride* and her short stories and essays; and E.D.E.N. Southworth's *The Hidden Hand* and her short stories. When I used these texts, I found students' translation activities more correctly reflected their English ability and I could evaluate each student's class activities more precisely.

Reading minor texts that many Japanese ESL students aren't familiar with does not give students a good historical perspective. These minor literary authors are not much mentioned in literary history and do not have introductive guidebooks for their study. Students cannot gain the proper historical insight to analyze the work they are translating. It is important to compare these minor works with other canonical works in order to give a proper historical position to the minor texts. This idea motivated me to compare W.C. Falkner's works with Faulkner's works in my ESL classes. Comparing these two Fa(u)lkners' works, I expected students to understand the literary development from W.C. Falkner to Faulkner.

This research was conducted in two classes. My undergraduate course, English and American Literature and Culture class (henceforth EALC), had eleven sophomores. In EALC, I compared W.C. Falkner's *The Spanish Heroine* with Faulkner's short story, "Wash." In my Seminar class (henceforth Seminar), which had fourteen juniors, students translated W.C. Falkner's *The Little Brick Church* and compared it with *Intruder in the Dust*. When Seminar students read *Intruder in the Dust*, they used the translation and the movie version to understand the story.[8] In these classes, both author's works were compared as examples of the two stages in Southern literature's historical development from romantic literature in the nineteenth century to modern literature in the twentieth century. As W.J. Cash suggests in *The Mind of the South*, Southern culture developed from the age of "the tendency toward unreality, toward romanticism" to the age of "the new Southern authors" that disdain their previous generations' sentimental atmosphere.[9] In these two classes I guide students to reexamine how Faulkner criticizes W.C. Falkner's nineteenth-century Southern romanticism.

These two classes focused on comparing two Fa(u)lkners' works to understand that W.C. Falkner wrote romantic stories that supported plantation culture in the antebellum South, while Faulkner overcame his great-grandfather's worldview with his modern works. Much of Southern popular literature from the nineteenth century to the early twentieth century nostalgically endorsed Southern plantation culture and slavery. In the early and middle of the twentieth century, Faulkner critically examined how modern Southerners encountered this regional legacy. If an ESL teacher wants to make students understand the historical development of the literature of the American South, it is beneficial to read W.C. Falkner's work as a typical example of the American South's romantic culture in the late nineteenth century.

In my two classes, on the one hand students translated W.C. Falkner's works into Japanese by themselves, but at the same time they used existent Japanese translations to understand Faulkner's writing and discuss with their classmates. Knowing the romantic pattern of W.C. Falkner's writing and its relationship to Faulkner increases students' understanding of the development of the form of literature within a region. The ultimate purpose of my classes was to create a dialogic space for ESL students' autonomous approach to a particular regions' literature.

I apply the concept of social combination in my pedagogy in order to create a dialogic space in ESL class, and to insist on the democratic aspect of American literature. In "The Power to Speak and Listen: Democratic Pedagogies for American Literature Classrooms" (2005), Robert Petrone and Robert Gibney argue that their pedagogy develops the hope that their American literature classroom will expand spaces for the multiplicity of voices that define democracy, and that their "aim is to create opportunities for democratic transformation for our students, their communities, and the world."[10] In the same way, my ultimate goal was to efficiently create opportunities for ESL students to experience the method for reading American literature and building democratic relationships among them.

Generally, in order to encourage students' democratic discussions, teachers usually use partner activities; however, I feel the concept of pair work does not create truly dialogical relationships between ESL students. The idea of pair work optimistically believes that the students can create equal conditions when the teacher sets them free; however, my experience tells me that students naturally create an unequal situation in class when the teacher unbridles students without any principles. In many cases, if a teacher asks students to find a partner for discussions, students tend to choose partners with similar cultural backgrounds. Many students believe that homogeneous pairs have a smoother discussion. This pattern can be found in all types of classes, not just in Japanese ESL classes. As Stephen Sharp writes in *A New Look at the Interactive Writing Classroom* (2011), in American ESL classes, students "like the fact that they have the power to choose their group members" and many instructors "prefer using the same groups throughout the semester," although these groups "can sometimes become too relaxed and friendly with each other, which can cause them to oversocialize and possibly lose their critical perspective."[11] Students show the same tendency in graduate classes in the United States. It is true that many graduate classes seem to have a rich multicultural condition; however, when I studied English and American literature at a graduate course in the United States, I found students tended to choose the same type of partner for their pair work. The pair's homogeneity was decided by various reasons: gender, race, language, nation, or region. In any case, the students kept the same partner for every session, and the class became the place that reproduced social gaps between different cultural groups. Japanese students at Seinan Gakuin University (SGU) were a relatively homogeneous group that shared the same language, and yet they were sensitive to their differences. Students decided their homogeneous partners according to gender, living area, pattern of life, hobby preference, and language level. Small differences in backgrounds increased the gaps between the groups.

Teachers should have some effective strategies for helping students communicate equally with their classmates. Quoting bell hooks's *Teaching to Transgress: Education as the Practice of Freedom*, Jessica Labbé argues that the classroom should be a place where teachers and students have an openness of mind and heart that allows us to imagine ways to "move beyond boundaries, to transgress."[12] Using small groups to discuss a particular theme and then changing up the members the next time is an effective way to help students have multiple communications with other class members. This is also an effective method to reduce the authority of the teacher. As Sharp suggests, in contrast to the knowledge-oriented method that strengthens the authority of the teacher, the small group activity shares the power, and when a teacher lets students work in small groups the classroom becomes "more of a democracy and less of a dictatorship."[13] This shift in power from the teacher to the students makes the class a more suitable place to create dialogic space in which students can offer their own points of view on various topics. In this dialogic space, both teachers and students can critically compare their viewpoints with others. In this activity, the teacher does not have the chance to lecture but instead guides students through a discussion.

If teachers are confident their classes can construct temporal, democratic relationships with each other, it is possible for them to choose any problematic text in class. For example, a teacher can use Thomas Dixon's *The Clansman* and D.W. Griffith's *Birth of the Nation* to discuss the history of racial segregation and its cultural formation in the American South. Or the other teacher can choose Earnest Hemingway's *The Sun also Rises* without worrying about the bad influence of the protagonist's anti-Semitism on the students. Whether it is a minor or popular text, democratic class activity can accept any problematic text and criticize it for egalitarian purposes.

Since Alexis de Tocqueville declared that "literature was an arsenal open to all from which the weak and the poor daily drew arms" in *Democracy in America* (1835),[14] American literature has been deeply associated with the process of the realization of democracy. As an American literary teacher of ESL students, I demanded to establish a democratic condition in class in order to educate each student as a good consumer and producer of various egalitarian behaviors and texts. As Petrone and Gibney conclude, teachers of American literature should "expand rather than restrict spaces for the multiplicity of voices that define democracy" because teachers hope today's ESL students who read American literature become "not only active and critical but also skilled consumers, producers, and distributors of texts and information."[15] I hoped the discussions and texts ESL students produced after reading American literature were less discriminatory and less dogmatic. To achieve this purpose, democratic relationships should be encouraged in class activities.

In my classes, the concept of social combination was used to connect ESL students with the study of the literature of the American South. The literature of the American South is a place where people personally struggle to have their own equal relationships to resist various powers that create social stratifications. Imagined expressions of people's struggles to have equal relationships in the American South are still worth studying for ESL students because students' personal connections, which are experiencing huge expansions through the digital network, reveal the world's deeply stratified conditions as well as global democratizing movements. ESL students can view the literature of the American South as a metaphor for today's world.

In the early stages of my courses, I told students that social combination had three features. First, social combination is an interpersonal communication among two or three people. Second, in social combination, the personal communication continues for a certain period of time. Third, social combination aims to attain an equal relationship among two or three people for as long as their relationship continues. I used these definitions to temporarily create egalitarian relationships among students. By applying the concept of social combination to ESL class discussions, students were able to have different partners every time. I hoped students would find fresh partners in each session to find new perspectives in their discussions. Without dialogue with these unfamiliar peers, the student does not understand how his/her own opinions and ideas are different from others. One of the most important purposes of social combination is students' awareness of the "otherness" of their class members.

The practical exercises of social combination in class are simple. After reading and translating American literary texts, students exchanged their opinions in two- or three-person groups in class and use the results of exchanged opinions and discussions to compose their digital writings on Moodle after class. In contrast to the ordinary pair work, students change their partner every session. I advise my students to give their best effort to have equal relationships with others while they have social combination activities in order to make students treat their ideas equally. The ideal goal is that every student makes social combinations with all the other class members before the end of the term. In this activity, I try to reconfigure students' personal relationships as a place to produce public opinions. As Venessa Paech says in her argument on online community discourse in today's globalized network, "A successful community is personal, but its consequences are social."[16] Personally knowing the other classmates' viewpoints is useful for students to criticize their opinions from many points of view. Students' dialogic activities help them compose their writing by reflecting on their partners' opinions. Through these activities, students experience how novels can be discussed with others and how their writing can contain plural perspectives.

The social combination activity has limitations on the size of group and length of time. A two- or three-person group is the smallest constitutional unit for a society, and it is also the best number for training in dialogue. The group's communication should be limited to five or ten minutes to attain the activity's practical purpose. If possible, teachers may allow more than ten minutes for discussion, but in undergraduate classes five to ten minutes is best. These limitations of group size and time help students understand equal relationships as practical actions.

If we do not consciously compose our equal relationships in our personal activities, we easily lose our needed efforts to build common ground in our personal relationships. As Hannah Arendt argues in *The Human Condition* (1958), in modern society, equality is a term that means an officially equalized system in public society, not equal conditions in personal actions: "society equalizes under all circumstances, and the victory of equality in the modern world is only the political and legal recognition of the fact that society has conquered the public realm, and that distinction and difference have become private matters of the individual."[17] Today we must recapture our sense of equality in our private actions. Democracy—officially understood as a political procedure of society—should also be reconnected to our personal actions. In *The Saturated Self* (1991), Gergen critically argues that democracy is a common matter but we have not seized it in our ordinary deeds: "We speak about democracy as an existing essence—a form of government to be studied by scholars, appreciated by citizens, and maintained by force if necessary. Yet the meaning of the term is not derived from observing persons moving about in time and space."[18] We still do not have proper activities or methods to seize our own personal realms in democratic ways. Social combination activities in class are practical ways to reconfigure democracy and equality as common customs in our interpersonal actions.

Along with social combination activities, I use translation activities to make sure of the correctness of ESL students' understandings. So far, in traditional ESL courses in Japan, English reading skills and comprehension have been strongly linked in Japanese language paraphrasing. Students take too much time in translation activities and not enough time comparing one literary work with another or contextualizing texts in cultural and historical perspectives. As Upton argues in "First and Second language Use in Reading Comprehension Strategies of Japanese ESL Students" (1997), ESL students "find it more comfortable to paraphrase or restate sentences and phrases" in their first language "just to confirm their comprehension."[19] As Seng and Hashim write, in general, teachers and ESL students can view translation as an important indicator of the development of ESL students' comprehension processes.[20] If a Japanese ESL teacher focuses on a close reading of a literary work, it means the teacher ensures student comprehension using a translation. Or, a Japanese ESL teacher may require students to sum up each paragraph in Japanese to make sure students understood the paragraph's plot or argument. In both cases, Japanese ESL classes have to use much of their time checking the meaning of English sentences—in many cases, the teacher requires students to translate or sum up an English text for their verbal presentation in Japanese. As translation activities consume huge hours for both students and the teacher, they do not have much time left for discussing what they are reading or what relationships they can find between texts.

To support students' social combination and translation activities in my classes, I use three digital platforms. First, I use Moodle to record students' paragraph writing as a result of their social combination activities in class. Two additional online platforms, WordPress and Google Drive, support students' understanding of English texts and their translations and enable collaborative improvement of literary translations. These three online platforms are used for the following practical purposes:

1. Moodle is a popular e-learning platform used in many Japanese universities. SGU's Information-Processing Center (henceforth IPC) uses Moodle on the university's intranet. I cooperate with IPC to create an original posting system, the Social Combination Sheet (henceforth SCS), on Moodle. This system is used by students to post their writings each week. Moodle is not applicable for the students' interactive and collaborative communications,[21] but it is useful for individual posting and teacher checking. I use Moodle so individual students can record their results of class discussions. Teachers can use Moodle's SCS to connect discussion activities in class and composition activities on the web. Even if teachers do not have Moodle in their academic institutions, MoodleCloud is now offering free hosting services that can be used by everyone.[22]
2. WordPress is a free blogging platform. I use WordPress to publish my model translation to support students' understanding of minor literary works that have not been translated into English. In their preparation activities, students translate W.C. Falkner's work by themselves to improve their translation abilities. The results of their translations are checked in class. After checking their translations, the teacher publishes the model translation on the Web because many students still do not have confidence in their own translations. WordPress is introduced as a platform mainly to make the teacher's information more accessible to students.
3. Google Drive is a free cloud storage in which people can use Word-style files (Documents), Excel-style files (Spreadsheets), PowerPoint-style files (Presentations), short question or mini-exam sheets (Forms), and easy

drawing sheets (Drawings) to use and share information with others. Students share their texts with the other class members and edit them collaboratively. Google Drive is a typical Web 2.0 platform that facilitates sharing and interacting with information. I use Spreadsheets as a question-and-answer sheet to instruct students on translation and use Documents so students can upload their translated texts and comment and edit them with the other students.

Moodle, WordPress, and Google Drive have their drawbacks, and it is important to use these online systems appropriately according to the activity's purpose. I created SCS on Moodle as an online posting system for students' English writing. WordPress supports students' translation activities, and Google Drive improves students' interactive activities with their peers and the teacher.

Students compose their opinions in English using SCS to improve their dialogic relationships in class and to sum up their discussions on the web. In Japanese ESL classrooms, it is easier for a teacher to develop students' discussions in Japanese in class and then have them summarize their results in English on Moodle. In this way, students compose their paragraphs including not only their own voices but also the voices of their peers. As Hirschkop argues, our intersubjective condition teaches us "the recognition of others in determining the meaning of our actions" and "our need and desire for narrative coherence and justification in our actions."[23] This sense of coexistence with others can be learned in this dialogic method.

The students' writing on SCS reflects their interactive voices in class. After every session, students review their arguments in class and write a paragraph on Moodle, showing their opinions in the right column and their partner's opinions in the left one. Then they write a paragraph in the middle column to integrate both opinions. Social combination activities and SCS posting on Moodle connect class discussion with writing activities on the web and help students understand their writing as the result of their dialogic communications.

The main difference between my two classes is whether I give students the model translation or not. In EALC, I use WordPress to offer a model translation in blog form. In Seminar, I use Google Drive to create interactive activities with students. In EALC, students take forty minutes each session to translate the original text of W.C. Falkner's *The Spanish Heroine*. After they finish their translation, I show them my model translation on WordPress. The students enjoy these translation activities as their linguistic exercise, and they expect they will find the model answer in my translation on WordPress. In this case, interactions both in-class and online are more teacher-oriented. In contrast, Seminar students translate the whole text following their own plan and they explain the plot of their translated part in every class; I do not show them the model translation. Then they use Google Drive to edit and comment on their translation by themselves. In this case, they feel pressure and frustration with the result of their translation activities. This frustration continues all year while they share their incomplete texts on Google Drive for their improvements.

The ways to compare both Fa(u)lkners' texts are different for these two classes. In EALC, each student writes a short summary of W.C. Falkner's *The Spanish Heroine* in Japanese every week and compares it with Faulkner's "Wash." They discuss the differences in these two works and post their opinions to Moodle. In Seminar, students translate all of W.C. Falkner's *The Little Brick Church* by themselves over the course of the year. While translating, Seminar students also compare *The Little Brick Church* with Faulkner's *Intruder in the Dust*. Students in both classes discuss Fa(u)lkners' works and post their opinions to Moodle.

The following reports are divided into four parts. In the first part, I explain how to use Moodle's SCS to help students compose a short paragraph that includes both their own and their partners' opinions. The second part reports on how students translate two of W.C. Falkner's literary works in class. In the third part, I outline how to teach American literature in ESL classes in more democratic ways, and argue how people can use today's digital network to build common ground with others. The fourth part suggests some ideas for future class activities.

SCS ON MOODLE

Moodle is a useful tool for students. As Yaman argues, "Moodle addresses to all educators and students without exception of any age group. . . . When we have a deeper look into the subject from this perspective, we can easily see that anyone who can handle basic computer operations will have no problem using Moodle."[24] So far, I have mainly used this platform to share class files and gather student writings, but in this report I explore how Moodle can be used as a prototype to help students develop their ability to reflect on their dialogic activities in their English writings. In class, my students develop their arguments in their social combination activities and then write their paragraph on SCS.

The purpose of the social combination activity in class and SCS on Moodle is to make students compose a paragraph in a dialogic condition with other classmates. SCS supports students in composing paragraphs in English after class discussion. This method's first stage includes having social combination activities in class—students read two Fa(u)lkners' works and compare their opinions on these texts. After they have read and translated their text, they discuss for ten minutes or so. In these discussions, students compare others' opinions with theirs. Then they try to find some terms or viewpoints that will integrate their opinions. If they find their opinions are the same, they will discuss why. Their dialogic activities in class are most often conducted in Japanese. They record their discussions on their paper, including (1) their own opinion, (2) their partners' opinions, and (3) the key terms or viewpoints to integrate them. In the second stage, after class, they compose their English reflection paragraph using SCS on Moodle.

SCS on Moodle is a prototype that makes students integrate plural voices in their paragraph writing. Today, every teacher can create this SCS system on Moodle for free using MoodleCloud.[25] In this system, a student uses three columns on the Moodle screen to compose a paragraph (see figure 3.1).

A student posts his/her first idea as Element A in the left column, their peer's idea as Element B (and also post the third opinion as Element C if they have the third member in their group) in the right column, and then the integrated paragraph—in which the student combines their opinion with their partner's opinion—in the middle column. These paragraphs are the result of student efforts to contain two opinions, Element A and Element B. This writing process demonstrates Bakhtin's so-called "double-voiced" condition. Bakhtin argues that "someone else's words introduced into our own speech inevitably assume a new (our own) interpretation and become subject to our evaluation of them; that is, they become double-voiced" (PDP 195). Using SCS, students systematically compose this double-voiced condition in their English writing, resonating their voices with their classmates' voices in their paragraph. What I want to create in student writing is this democratically arranged double-voiced condition.

SCS on Moodle improves students' dialogical relationships on the web. Using Moodle for student composition activities may not sound interactive, but I find this system can increase students' dialogues with others as well as their creativities in their digital writings. SCS on Moodle encourages students to compose their own voice by reflecting on other students' voices.

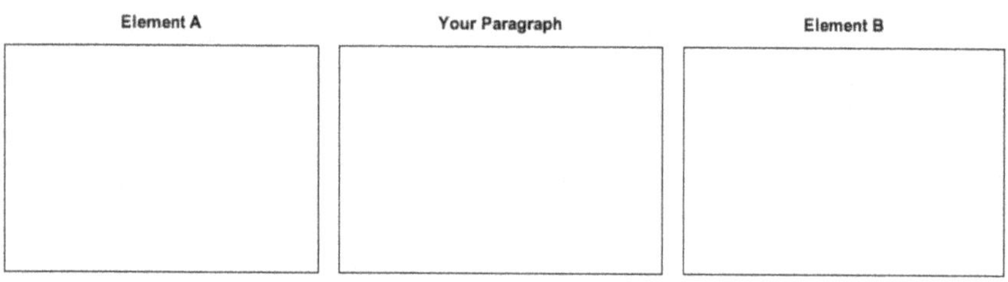

Figure 3.1 A sample of three columns.

A student may create his/her own paragraph, but that paragraph still shows the writer's effort to reflect their partner's opinions. Teachers can evaluate a student's paragraph not only as a result of individual effort but also as a result of the social combination activity in class. The teacher can easily check the students' combined opinions and track their original ideas by checking Element A and Element B. This rigid system is more useful for teachers to check how students interact with their writing than other SNS platforms. Although some scholars argue that SNS, such as Facebook, offers a space where "the relatively controlled academic discourse and interpersonal hybrid and informal discourse styles" meet,[26] in many cases students are only writing short sentences on these SNS that do not include other students' voices. It might be true that students can use SNS for their communication, but I argue SNS writing does not improve students' ability to compose their writing dialogically. In many cases, they concentrate more on other members' reactions. When they post photos or short comments, others push Facebook's "Like" buttons to respond positively to the uploads, but they do not have many chances to hybridize their words with others' utterances. Students are absorbed in Facebook and other SNS as if obsessed with this action and reaction game. Some conversations might be exchanged in the comment columns, but their conversations do not often develop into Bakhtin's double-voiced condition. Even today, if a person wants to create a double-voiced condition in his/her own utterance, he/she has to consciously compose every paragraph.

Prior to my introduction of the method of social combination and SCS in class, I could not connect students' class discussions with their writing activities. When I asked students to submit comments in English on the text or on their discussion, many students wrote two or three short sentences without composing a full paragraph. Often, student writings after class discussions did not reflect their discussion experiences in class. In contrast, SCS helps students compose a paragraph that is connected to their class discussions.

My social combination activity follows four steps to develop an argument. First, students write short reflections on what they think about the day's reading or translation activities. They write at least one sentence in English or in Japanese. Second, students form two- or three-person groups and they exchange their written materials with their partner and read each other's opinions. Then they give some responses on their partner's ideas. Third, students compare their opinions with each other and find and write the common element, idea, or concept that bridges their opinions—this might be the key word of their topic sentence. Fourth, students return their papers to the original writers and then write their new paragraph. They can use their agreed element, idea, or concept that bridges their opinions as a topic sentence for their paragraph, and then use Element A and Element B (and Element C) to support or examine their topic in various points of view. They are suggested to add their conclusion viewing their topic in different light. Following this procedure, they complete one paragraph. This paragraph can be written in Japanese. If possible, the teacher should look over their writing and give advice since students may not understand how to connect two or three opinions in a paragraph.

After finishing their short discussions in class, students are then asked to write their paragraph using SCS on Moodle. Students should complete their paragraph by themselves after class by remembering their discussion and recomposing the other students' voices in their own English paragraph. When they open SCS on Moodle, students find the following guideline on their monitor (see figure 3.2).

The goal of their paragraph is to create a persuasive argument that includes both points of view. The following guideline is the detail I wrote for SCS assignment:

1. Review your class paper in order to write you and your partner's opinions on SCS. Integrate the paper's elements to write your new paragraph in English. You may have already written the keywords to bridge your opinions on your paper.
2. Enter Moodle's SCS site and write your opinion on Element A, and then write your partner's opinion on Element B. If you have two partners, write your two partners' opinions in a row on Element B. Both columns should be written in English.
3. If you have not already discussed the idea or key terms that bridge Element A and Element B, consider the idea that will bridge or integrate the two elements. This idea can also be developed into the topic sentence of your paragraph.
4. Consider the last sentence that will wrap up the whole paragraph. You can repeat your idea you have already said in the topic sentence, but try to express it from a different point of view or write it using different words. A thesaurus dictionary is useful to find another way of saying.
5. Create a connecting statement that will smoothly join Element A and Element B. The model structure of your paragraph will be this: Topic

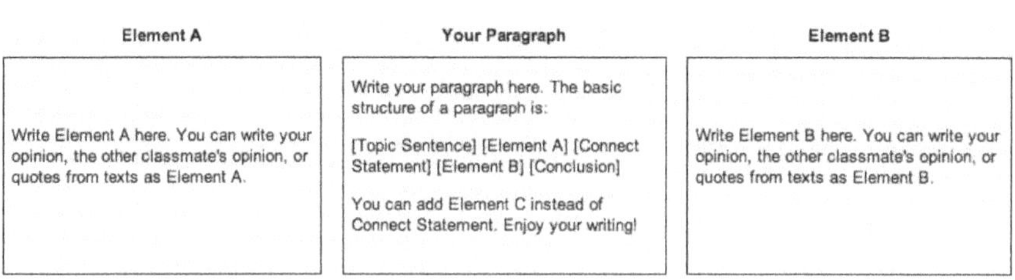

Figure 3.2 SCS guideline.

Sentence / Element A / Connect Statement / Element B / Conclusion. You can add Connect Statement / Element C after Element B if you have the third partner's opinion. Read through your paragraph and adjust some words to smoothly connect every element to complete the logical connections.

Students follow these five steps to complete their paragraphs, and I check their written paragraphs on Moodle. I can see what elements the students used and how they developed their paragraphs. Some students may not complete the whole task on SCS, but the blank columns are obvious and it is easy for the teacher to check and encourage them to fill in all three columns. In my experience, Moodle made it easy for the teacher to see and check their students' writing, and in many cases students improved their writing in each SCS session.

Sometimes I was afraid this system was too stressful for students because some students had reacted negatively to English composition. However, I found that almost all students could positively take part in this activity. In the social combination activity, students enjoyed personal exchanges of opinions in class and then they could connect their discussion with their writings on Moodle to open up their English writings to their classmates. Students seemed to write their paragraphs earnestly because of their awareness of its public quality. In the next part, I explain what kinds of mutual and dialogical effects SCS system had on student writing, including some samples from both classes.

Effects of SCS on Student Writing

In SCS on Moodle, students' writings in a session are juxtaposed on one site page and they can see how their peers compose their opinions. This open space increases students' dialogic relation and builds common ground among their writings. They begin composing their paragraphs while still being aware of their peers' writings. At the same time, they can see how their partner reflects their voice in their own paragraph and combines plural opinions or perspectives in one paragraph. In the following example, Student X and Student Y had a social combination activity in class. After class, they completed their paragraphs on SCS, but they were able to look at each other's paragraphs and see how their partner expressed their opinions on the same Moodle screen. In this case, Student X and Student Y learned each other's opinions in class and can check each other's writings to see how their partners expressed their opinions (see figure 3.3).

The teacher could see that Student X's opinion (Student X's Element A), "In Intruder in the Dust, white people, especially white men have a tendency

Element A	Your Paragraph	Element B
In Intruder in the Dust, white people, especially white men have a tendency to look down on African-American people unconsciously.	"White people tend to look down on African-American people" is our topic and there are two reasons why we chose the theme. Firstly, in Intruder in the Dust, white people, especially white men have a tendency to look down on African-American people unconsciously. For instance, a white man in this movie calls African-American men "boys" and it intends to demonstrate a disdain for them. Secondly, white people are liable to treat African-American people like nuisances. It is said that white people put lynching of African-American people on show at that time, though a discrimination of African-American people has been already abolished. Thus, we can see from this movie that there is still a discrimination of African-American people, and white people must have a condescending attitude toward them.	white people are liable to treat African-American people like nuisances.

Student X

Element A	Your Paragraph	Element B
The white man looks down on the black unconsciously.	White people have hidden contempt for black people. First, the white man looks down on the black unconsciously. He pretend to act as equals, so this rank relationship is not obvious at first glance. In addition, white men assault the black in a group at night, when few people can see their violation. In conclusion, white man doesn't want the public to know he oppress black people.	White people assault the black in a group.

Student Y

Figure 3.3 Student X and Student Y: Oct 8, 2014.

to look down on African-American people unconsciously," was simplified in Student Y's Element B: "White people assault the black in a group." In the same way, Student Y's opinion (Student Y's Element A), "the white man looks down on black unconsciously" was reflected in Student X's Element B: "white people are liable to treat African-American people like nuisances." In these mutual reflections, students saw how their comments were expressed in their partner's words. Their paragraphs showed how they exchanged their opinions in class and then analyzed their partner's opinions in their own writing. These columns suggest SCS is a useful tool to make students interpret and re-accentuate their partner's voices in their paragraphs. I find that students sometimes use inappropriate terms and make grammatical errors, but because of their collaborations their writings become more social and easy to read. The students' dialogic writings are more understandable and interesting than their monologic writings. This is one of the good effects of SCS system on writing.

When students write on SCS, they are aware of their classmate's voices. When they reflect on their partner's voices in their paragraphs, their paragraphs become more plain and persuasive. This process teaches students what kind of quality of writing is needed for their contemporary audiences. When students communicate with their partners, they find their ideas have an audience in class and their writing changes from self-oriented monologue to dialogical discourse. Their writings express their understandings of the other class member's opinions and desire their classmates' comprehensions. ESL students need to reconnect their arguments with other class member's opinions in their English writing activities. Even if these students discuss in Japanese, it is useful to know plural viewpoints on their texts and then to gain a more general idea of how to integrate multiple opinions. The partners offer new and different perspectives, then their Japanese discussions will be reflected in their writings in English. Thus, the combination of multiple voices becomes more powerful and communicative. As Oster argues in "Seeing with Different Eyes: Another View of Literature in the ESL Class" (1989), it is possible to use literature to realize "one's own cultural blind spots when one compares one's own responses to a literary text with someone else's."[27] Conversations with others on a literary text help students recognize other viewpoints. In the following example, a student found a new perspective by comparing two opinions about *Intruder in the Dust* (see figure 3.4).

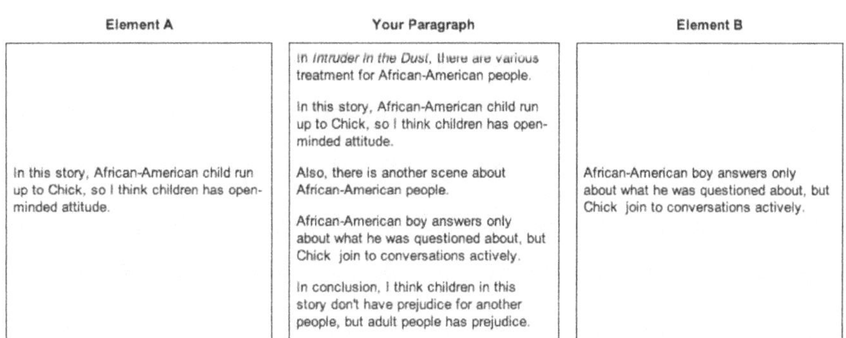

Figure 3.4 Student Z: Oct 2, 2014.

In Element A, a student described her opinion that Chick's African American partner, Aleck Sander, seemed to have an "open-minded attitude" to Chick. Element B showed that this student's partner found a difference in the attitudes between Aleck Sander and Chick. In front of the white adults, Aleck Sander keeps quiet and answers only when he is questioned, while Chick can take part in the adult conversation "actively." In her own paragraph in the middle column, Student Z analyzed this difference and concluded that Chick does not have a "prejudice" against Aleck Sander, but the white adults do. The middle column was the result of the student's effort to integrate two different viewpoints to make her argument persuasive.

I use SCS in order to give students a way to structure a paragraph. In a usual writing system, the ESL teacher introduces the paragraph model with a topic sentence, some supporting sentences, and a conclusion. Students can compose their own paragraph following the model, but often students do not recognize that writing a paragraph is the result of discussions or dialogue between two or three individuals. In the first stage of my class, students did not have any method for connecting their dialogical experiences to their paragraph. However, when I taught them to connect two or three students' opinions in a paragraph with common key words, they could develop their own paragraph using plural viewpoints after their discussion. Using SCS on Moodle, students realized their paragraph structures stem from their discussion activities in class. In the early stage of SCS activity and writing, students just react to the teacher's requirement. In the later stage, I led students to compare two Fa(u)lkners' works and to develop their own discussions to complete their paragraph. The following examples show how students develop their writing abilities using SCS on Moodle.

SIX RESULTS OF STUDENT WRITING IN EALC AND SEMINAR

The following six samples show how student compositions develop over a month. In each case, the first example is a student's early writing and the second example is written a month later. In EALC, students answered some general questions about learning English and American literature in their early SCS writings, and then they argued about two Fa(u)lkners' texts in their later writings. In this class's case, the comparison between the students' early writings and later writings may not precisely prove the students' development of ability to write their arguments in their paragraph because their writings' purposes are different between the early versions and the later versions, but still I feel these comparisons suggest the development of students' consciousness on their English writings and literature readings. In

Seminar, students compared the two Fa(u)lkners' works from the start of the course to the end. In this case, their writings' themes constantly mentioned the comparisons of the two Fa(u)lkners' works, and the progresses of their writings could be appropriately examined comparing their writings' early versions with the later ones. All these students submitted their paragraphs once a week—this meant that they posted their writings three or four times between these two samples. Their development showed how student writing could improve after three or four tries. In EALC, samples were collected during the first term in 2014, and in Seminar, samples were collected from the first and second terms in 2014.

Student A in EALC

In the early stages of EALC, Student A wrote a stereotypical response to my question about the meaning of English study, but after he read *The Spanish Heroine* he found that W.C. Falkner's novel was more interesting than Faulkner's short story, "Wash." He explained his preference in his later

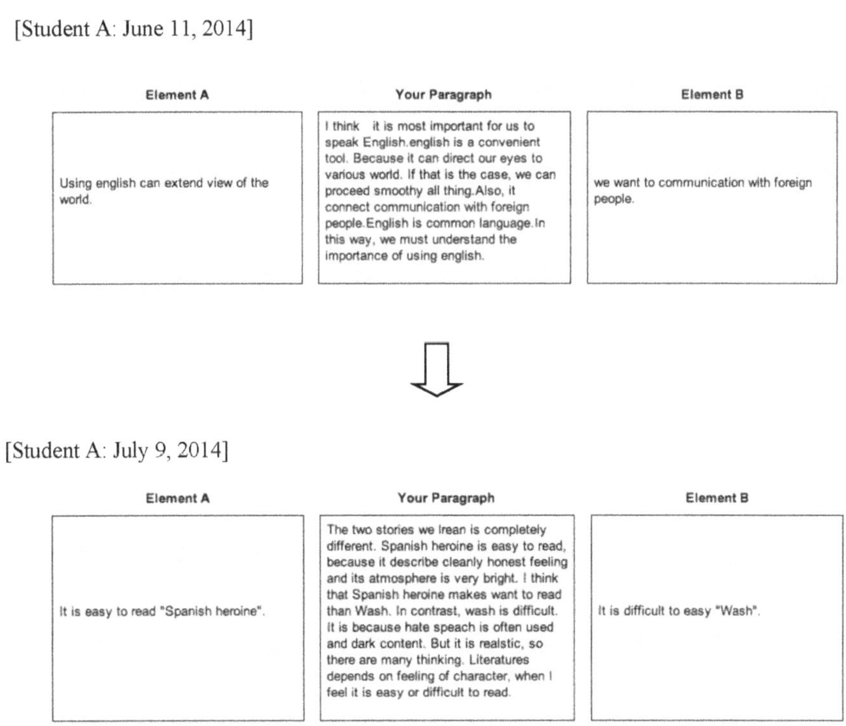

Figure 3.5 Student A.

paragraph. Compared to the early one, his later writing focused on his own personal opinion of a particular work, and his argument reflected his own voice and the other class member's voice more persuasively (see figure 3.5).

In his later paragraph, Student A explained why he liked W.C. Falkner's *The Spanish Heroine*. He said *The Spanish Heroine* was easy to read because "it describe cleanly honest feeling." In spite of many grammatical and word-choice mistakes, his later writing was more attractive. He had discovered his literary preferences and wanted to express his honest impression. Comparing *The Spanish Heroine* with "Wash," he understood that readers of "Wash" could not expect the typical "heroic" element from *The Spanish Heroine*. Recognizing that "Wash" is "realistic" and its atmosphere is "dark," and that *The Spanish Heroine* is "easy to read" and "its atmosphere is very bright," Student A found he was more attracted to *The Spanish Heroine*. This recognition of his own preference makes his writing more evocative and enticing.

Student B in EALC

Some students did not show much improvement on their paragraph structure. These students had a relatively high ability to compose English from the start, and so they stuck to their style and did not develop their ability to

[Student B: June 1, 2014]

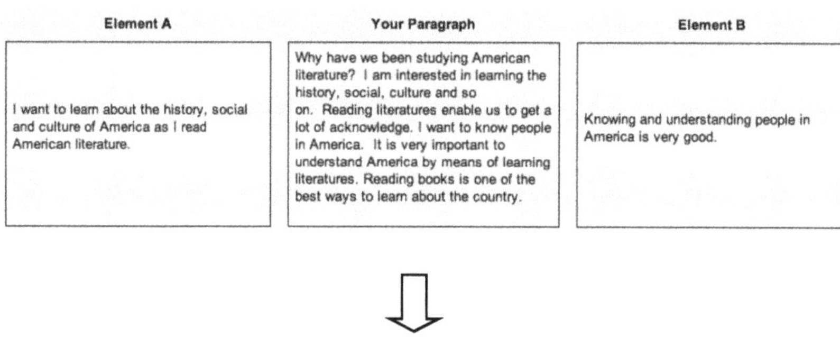

[Student B: July 2, 2014]

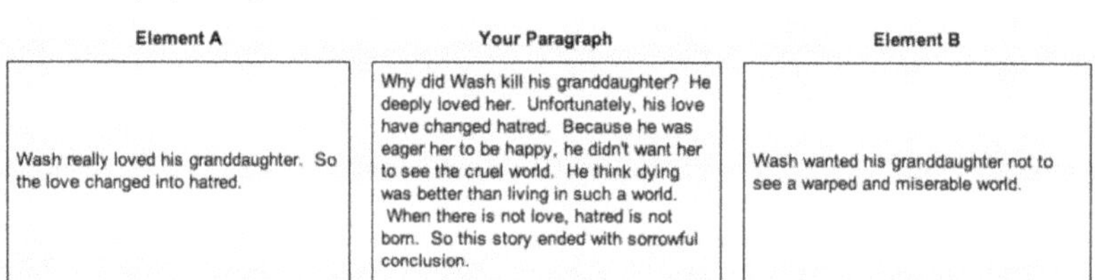

Figure 3.6 Student B.

compose their logic. Still, SCS activity gave them chances to develop more specific details. Their writing went from general statements to more delicate arguments. In Student B's case, her early writing showed a good sense of English, but it was a typical, general statement; it arranged some stereotyped opinions on American literature, and her writing did not reflect her own voice. However, one month later, the same student seemed to find her own opinion on Faulkner's short story, "Wash." She tried to explain why the protagonist, Wash, killed his granddaughter. As she analyzed the protagonist's psychological condition, she developed her argument using cause and effect and her supporting sentences showed logical coherence with an emotional understanding of human nature (see figure 3.6).

In her later writing, Student B analyzed why Wash changed his mind and killed his granddaughter. She guessed that Wash thought "dying was better than living in such a world" and therefore "this story ended with sorrowful conclusion." If she adds some quotes from Faulkner's short story to her later paragraph, it will more thoroughly analyze Wash's reason for killing his granddaughter. Her writing shows her ability to expand her opinion by focusing on one theme.

Student C in EALC

In her early paragraph, Student C used an immature structure to answer one of my questions—what is literature. She wrote a generalized statement on literature and did not show any plural viewpoints on the topic. In her later one, however, she developed her own idea in her paragraph, though the sentences were arranged in almost the same way as the early paragraph (see figure 3.7).

In her later example, Student C showed the ability to develop a logical argument to explain her own question. In this case, her writing ability became more specific and focused on a particular author's work. Student C's early post showed enough effort to give a good answer to my question, but a month later she wrote her own opinion on Faulkner's "Wash." She argued, "Wash is antihero" because Faulkner "expressed white trash's suffering at that time through Wash." She compared her opinion with her partner's argument, which insisted "Wash is hero" because Faulkner described Wash as a "hero for poor white at that time." To integrate these antagonistic opinions, Student C used "reality" as a key word. In her conclusion, Student C argued Wash did not have "heroic elements" because "Faulkner featured Wash" in order "to describe reality by focusing on poor white." Student C effectively developed her argument by focusing on the purpose of the story.

[Student C: June 1, 2014]

Element A	Your Paragraph	Element B
Literature is like a mirror of the reality.	What is literature? It is like a mirror of the reality. Because literature works like mirror, we can say as following. It can be changed by historical background, but the quality of literature doesn't change. So, literature reflects the writer and historical background by working like a mirror, and it is also tool to convey something.	The quality of the literature does not change but the content can be changed by historical background or matters at that time.

[Student C: July 4, 2014]

Element A	Your Paragraph	Element B
Wash is antihero. Because Wash is white trash, Faulkner expressed white trash's suffering at that time through Wash.	Is Wash hero or antihero? Wash is antihero. Because Wash is white trash, Faulkner expressed white trash's suffering at that time through Wash. On the other hand, we can regard Wash as hero. Faulkner described Wash as an example of poor white. Therefore we can argue that Wash is hero for poor white at that time. He is lack of heroic elements. The reason why Faulkner featured Wash was to describe reality by focusing on poor white.	Faulkner described Wash as an example of poor white. Therefore we can argue that Wash is hero for poor white at that time.

Figure 3.7 Student C.

Student D in Seminar

In the following three examples, Student D, Student E, and Student F show their SCS writings in Seminar. Seminar students had class sessions once a week. Students spent much of their time translating *The Little Brick Church* and did not spare much time for paragraph writing. Their SCS activities began at the end of spring term in 2014 and they continued their paragraph writing in the next fall term. I collected the following data from July 2014 to October 2014. Some of the samples show that students' English writing abilities developed in just two or three weeks. Since the students watched the movie version of *Intruder in the Dust* to help them understand the story of Faulkner's complicated novel, some students referred to the movie in their paragraph writing.

In his early stage, Student D did not follow the rule of SCS and write one sentence in each column. He did not have much interest in writing, but two weeks later he seemed to be fascinated by the story of *Intruder in the Dust* and he began to focus on the social structure of the South. Compared to his early writing, his second paragraph effectively analyzed the differences between people's reactions toward lynching according to their class differences (see figure 3.8).

Comparing to the early writing, the later paragraph of Student D showed the writer's will to compose his own opinion on the small community's stratified social structure. In the first sentence of his later writing, he suggested the lower-class white was frustrated with society and demanded to lynch Lucas. Then, he paraphrased Element B's short comment, "Over middle-class people do not join lynch[ing]," in order to write his own argument: "On the other hand, middle or upper-class people are not interested in this lynch because they were satisfied with the present situation." He rewrote his partner's opinion to refer to the Southern whites' stratified society and their different attitudes toward lynching. His analysis concluded that the

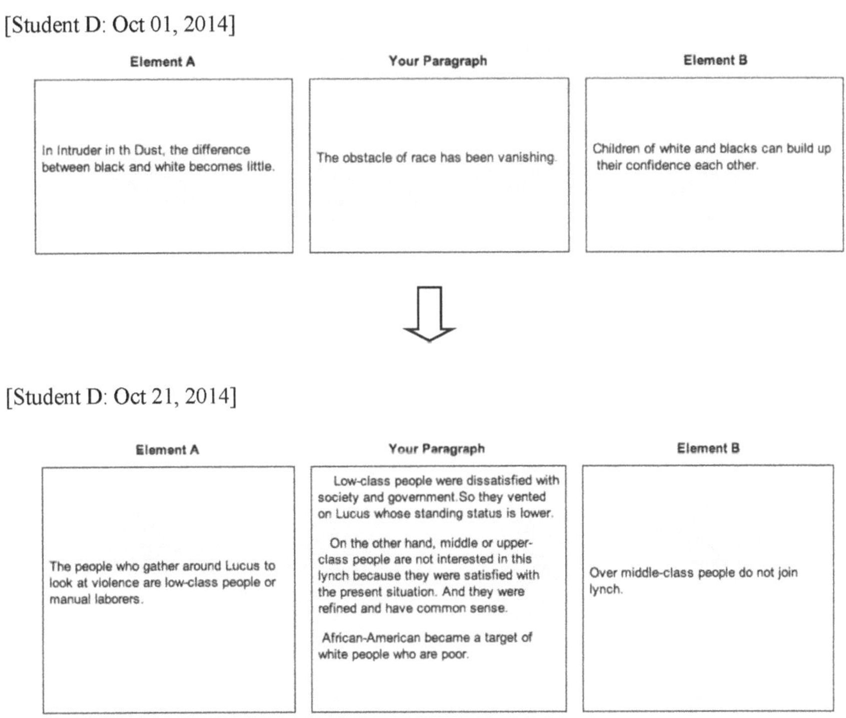

Figure 3.8 Student D.

lynching of African Americans was a vent for the poor whites' frustration. The later writing of Student D had a good logical structure and offered an interesting analysis.

Student E in Seminar

Student E argued about *The Little Brick Church* in the first sample. Her English was good early on, but she did not arrive at the dense analysis of a character's mind. She explained a hero's romantic relationship with his heroine and concludes that their love was an "important element to encourage us to be more interest[ed] in this story." This conclusion simply appreciated the romantic plot as an element to attract readers, but it did not critically

[Student E: July 24, 2014]

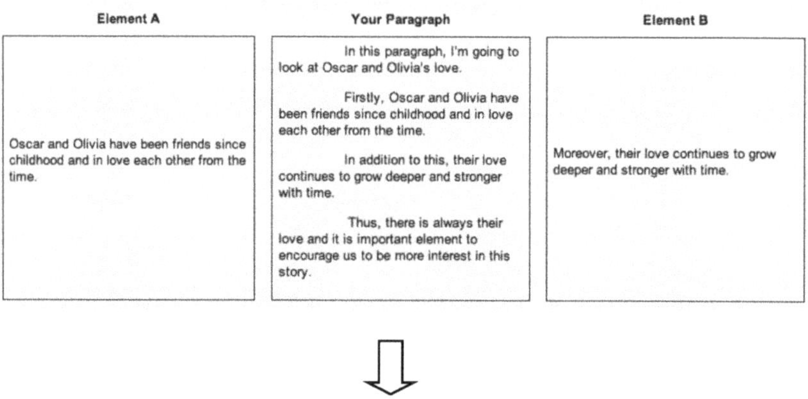

[Student E: Oct 15, 2014]

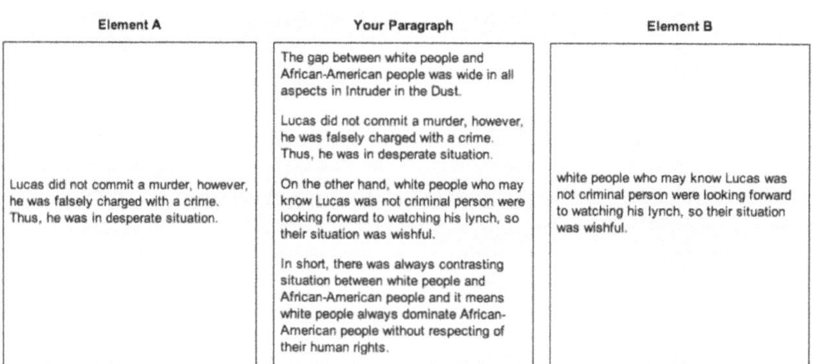

Figure 3.9 Student E.

examine the story from several points of view. In contrast, in her later writing, Student E examined *Intruder in the Dust* and focused on the "gap" between the white people and African Americans (see figure 3.9).

In contrast to her early writing, Student E's later writing showed her recognition of the society's cruel treatment of the victimized African American. The second sentence described Lucas's situation and the third sentence explained why the white people gathered. In the conclusion, she developed the third sentence and argued that "there was always contrasting situation between white people and African-American people and it means white people always dominate African-American people without respecting of their human rights." This conclusion can potentially be developed into a several-paragraph argument. Her topic sentence—"The gap between white people and African-American people was wide in all aspects in *Intruder in the Dust*"—was a nice summary of the whole paragraph. Element A and Element B were short sketches of her discussion with her partner, but Student E's paragraphs showed her ability to rewrite grammatical errors and make more sense. She integrated plural viewpoints in one paragraph to analyze the social structure and people's minds in *Intruder in the Dust*.

Interestingly, in her later writing, her own opinion, Element A, was not fully developed in her final paragraph argument. Instead, in the latter half of her paragraph, where she tried to reflect her partner's opinion, she developed her partner's point further. This suggests that some students show their English writing ability not when they express their own opinions but when they want to correctly express their partner's opinions. One of the reasons for this phenomenon is that some students find more unknown ideas in their partners' opinions and then they need much more explanations or developments to complete their writings. This interesting result suggests the importance for students to communicate with others to learn how to objectively compose their own arguments in English. Some students learn the way to develop their logic in a foreign language by communicating with others and rewriting their partners' opinions in English, and then they gradually learn how to objectify their own ideas and arguments.

Student F in Seminar

Students' early writings sometimes showed they had difficulties composing the topic sentence to integrate the opinions they recorded in Element A and Element B. Student F did not comprehend how to focus on her paragraph's topic in her early writing. When she wrote, "*Intruder in the Dust* describes people who were not often paid attention to before," her sentence did not

[Student F: Oct 2, 2014]

Element A	Your Paragraph	Element B
There are descriptions about the boy's worry and decision, while child is only an onlooker in *the Little Brick Church*.	*Intruder in the Dust* describes people who were not often paid attention to before. In this movie, the boy is troubled about Lucas and make a decision, while in *the Little Brick Church* child just stood by watching. Furthermore, there is another important scene. White people act for Lucas, who is a black. the black was treated as slave in *the Little Brick Church*. In conclusion, this story describes the child and black people who didn't establish independence yet. It focuses on thier individuality.	It is a different point from *the Little Brick Church* that white people act for the black.

[Student F: Oct 16, 2014]

Element A	Your Paragraph	Element B
Chick is the key of this story.	White people are at the center of things. First, Chick, a white boy, is the key of this story. His action and words effect on adults around him and develop the plot. There is another description which shows white people are in the lead. Next, mob violence is an amusement for white people. In other words, white people can make this situation pleasure without regard to Lucas' despair. In conslusion, things are mainly affected by white people's behavior.	Mob violence is an amusement for white people.

Figure 3.10 Student F.

integrate Element A and Element B. But two weeks later, her topic sentence was more developed (see figure 3.10).

In her later writing, Student F wrote a more concise topic sentence. When she wrote "White people are at the center of things," it was clear that she wanted to argue white people are in the dominant position in the community. The following elements supported this topic sentence enough to keep the whole structure of the paragraph. Though the later paragraph was shorter than her early writing, she could concentrate on her own topic and express what she wanted to say more concisely.

Summary: SCS in EALC and Seminar

SCS writing activity on Moodle is beneficial both for students and teachers because its columns clearly display the process of development in each student's paragraph. A teacher can compare Element A, Element B, and the middle column to see how a student rearranged and developed the elements to compose a paragraph. The teacher can also check whether or not the students created new ideas to integrate Element A and Element B in their topic sentences. As Sharp argues, there are "many educational benefits of teaching writing as a process" because teachers can "demystify" the writing process for students.[28] In my case, I wanted to teach my students the process to develop their own ideas through their discussions with other classmates. By connecting their class discussion with their writing, they could compose the dialogic condition in their writing.

Once every student wrote a paragraph after the session, I reviewed students' paragraphs in the next class session. Students' paragraphs were projected on screen and I briefly commented on all their writing. I found this review was a good way to develop students' writing abilities. I did not give detailed error corrections for their writing. As some ESL teachers suggest, error corrections can have a negative effect on students' motivation to write in English.[29] I instead gave some suggestions or model sentences to improve their writing, but I did not write my corrections on the board or on the comment column of Moodle to avoid copying. I asked my students to recompose my oral advice in their own words, which seemed to be more effective in improving their autonomous writing abilities. This SCS activity helped students connect their writings' class reviews with their next writings.

When I designed SCS on Moodle, I wanted to use an interactive platform like Google Drive's document or spreadsheet format. My first plan was to make an interactive chat system in which a person's opinion appeared on the left side of the Social Combination Sheet, and the other person's opinion would appear on the right side of the sheet, for students to cooperatively write the middle column by hybridizing two students' opinions. But as this plan was very difficult to realize on an ordinary digital platform, I decided to have students individually post their writing on Moodle's webpage. This composition system of SCS on Moodle is simple but easily guides students in composing their own paragraph using the other class member's opinions. Writers should check their opinions from different points of view so they can see their arguments from different layers and aspects. As Bakhtin argues, in the language diversity "a prose writer can distance himself from the language of his own work, while at the same time distancing himself, in varying degrees, from the different layers and aspects of the work" (DN 299). This argument is applicable to ordinary students' writing activities as well. Writers should take a proper distance from their own writings juxtaposing their voices with other people's ones. In SCS activity, by including plural voices in one paragraph, students can learn how to take a proper distance from their own opinions.

The purpose of this SCS activity is to bridge students' in-class discussions and their writing activities on the web, but this is an introductive activity for students to make dialogic relationships with other viewpoints. I hope students will expand their interactive relationships beyond their classmates' voices to include other texts and develop their academic writing. At the more advanced level of these activities, student's writings can incorporate other scholars' opinions. The teacher can instruct students to quote the scholar's opinions in Element B and develop those arguments in their paragraph. If they want to use several scholars' opinions in their argument, the teacher can instruct them how to treat these scholar's opinions as Element A and Element B—and possibly Element C, Element D, etc.—and connect them with their own ideas. The teacher can use this system for in-class discussions, online composition, and even in the first step of academic writing.

This idealistic development is difficult to achieve in my daily exercises. In many cases, students' writing suggested they did not follow the process I ordered. They frequently skipped some steps. Sometimes they did not write Element B. It is possible that students composed both Element B and Element A by themselves, not reflecting on their partners' opinions. Some students did not write topic sentences or conclusions for their paragraphs. However, the main purpose of SCS activity is to make students aware of the other class members' opinions when they write their own paragraphs. Students had many chances to discuss with their classmates and realize their paragraph is the result of multiple opinions and data. Students experience SCS system in a process from in-class dialogues to their composition of paragraphs.

Creating students' own original opinions using SCS system has three advantages. First, SCS activity prevents plagiarism—students must show what elements they used to compose their paragraph. In many cases, these elements are the result of discussions I have already checked. In this process, there is almost no way to copy other Internet sources or other writers' opinions. They have to show their own ideas and opinions in two element columns, and then their creative ideas should be combined into one paragraph in the middle column. This structure prevents students from directly copying some from the web and submitting it to the teacher. The second advantage of SCS activity is that it increases the students' originality and creativity. The teacher can instruct students to develop their own argument by combining their opinions with their partners' ones and creating their own worldview through dialogue. Students' opinions should be integrated with others to gain a broader perspective. Composition using SCS activity makes students aware of this development of their ideas. The third advantage is that students experience a sense of invention of their ideas in their dialogic experiences with other students. Students who come from various socioeconomic and cultural backgrounds try to have temporal, equal relationships with each other to create new ideas. The SCS activity is useful to find common ground for the students' inventions. Teaching social combination using the digital platform can help students be an original creator of their own ideas and express their inventions in their writing on the digital network.

TRANSLATIONS IN EALC AND SEMINAR

Translation is an important activity to make sure of the semantic connections between Japanese and English. My students translated two of W.C. Falkner's literary works in class. In EALC, students translated *The Spanish Heroine*, and in Seminar, *The Little Brick Church*. W.C. Falkner's works have not been translated into Japanese language, so my students had to translate the stories by themselves and compare them with Faulkner's literary works.

Many students were fond of W.C. Falkner's works. His works are full of heroic and brave scenes. His English is relatively simple and directly expresses the author's opinions. The endings are very clear and include a happy-ever-after. Students enjoyed them as if they were a weekly popular TV drama. Moreover, at the end of the term, they understood how these works' romantic narratives support white supremacy, masculinity, and class consciousness. Some students correctly guessed that W.C. Falkner's works are criticized in Faulkner's modern works. In my two classes, students experienced W.C. Falkner's works through their translation activities and studied Faulkner's works to understand how the two authors are connected.

Translation Activity in EALC

In EALC, students had three tasks: (1) a translation test of W.C. Falkner's *The Spanish Heroine*; (2) reading Faulkner's short story, "Wash," in Japanese; (3) a social combination activity to compare these two works. Since the class met twice a week, the students translated *The Spanish Heroine* in the first session of the week and then read and discussed Faulkner's "Wash" with social combination activities in the second session. They also wrote their paragraphs on SCS after they finished their sessions every week. Here I mainly focus on students' translation activities in EALC. The students read ten pages of *The Spanish Heroine* in English before class and wrote a summary in class for forty minutes. Near the end of this forty-minute translation test I checked their translation and gave between one to three points.

The class activity started with the translation examination. I gave them a blank paper and asked them to sum up ten pages of *The Spanish Heroine* in Japanese, translating some important sentences. Some students had already looked up the unknown words for their test in class, but others did not prepare and consequently did not achieve good results in their translation summaries. I walked around while they wrote their answers and checked them. The students' results obviously showed whether or not they had prepared for class. The students showed their various linguistic abilities and vocabulary knowledge in their translations. Grammatical accuracy also varied between students. Some students were good at summarizing while others were good at translating.

After they completed their summary of the story, I asked them some questions to make sure they understood the storyline. Then I gave them my model translation on WordPress. The model translation of *The Spanish Heroine* helped the students enjoy W.C. Falkner's story in Japanese. Their translation abilities were examined and graded every week, but their translations were far from the quality needed to enjoy the story. They still required the model translation for their enjoyable reading of the novel. This model translation supported their understanding of the story, and almost all my students enjoyed this model translation after they struggled to translate W.C. Falkner's works. Students checked the model translation as a review of their study. After translating English texts by themselves and comparing their translations with my model, students formed two-or three-person groups and exchanged their opinions on what they read that week and compared "Wash" with *The Spanish Heroine*. Through these activities, students were able to understand how these two author's works can be compared.

In this class, I used WordPress to give the model translation to my students. Students comfortably used this digital platform to compare the model translation with their own understandings. Additionally, in the social combination activity and writing on SCS, students experienced dialogic relationships and writings. They check their translations using WordPress, and then they have dialogic relationships with each other through the social combination activity and paragraph writing. These activities seemed appropriate for my students because of the balance between the teacher-oriented activity and the student-oriented activities in class.

Translation Activity in Seminar with Google Drive's Spreadsheet

In Seminar, the practical procedures of translation activities followed three steps: (1) allocate translation parts to students; (2) check the progress of translation tasks every week; (3) give teacher's guides for their translations using Google Drive's spreadsheet. These three steps helped each student complete their first draft of translation. The text in Seminar was W.C. Falkner's *The Little Brick Church*. Along with their translation activities, they had social combination activities and SCS writings to compare *The Little Brick Church* with *Intruder in the Dust*, as I have already reported students' samples in SCS on Moodle. The following process shows how Seminar students translated the 429-page novel in one semester.

First, I made four translation groups in class. As *The Little Brick Church* is in the public domain, I physically copied the whole book and divided it into fourteen parts. Some parts were around forty pages while others were closer to thirty depending on chapter divisions. I distributed these divided copies to each group. For example, the first group should translate page 1–167 and had four divided parts of the novel: page 1–37, 38–78, 79–134, and 135–167. The other groups were assigned parts in the same way. Then the members chose which part they wanted to translate in their group. I gave them the schedule sheet, which broke up their translations into twenty sessions. Students enjoyed their first encounter with their text in the first session. They compared their text materials with each other, tried to guess the story, and then decided which part would be the best for them. This process was important because the members recognized they chose their part by themselves and felt it was their responsibility to complete their translation.

Second, I asked them to decide on their group's student manager, who was in charge of keeping the schedule. In the following session, they were asked to report on the result of their translation and write a summary of the story they translated. This reporting activity was effective in connecting students' tasks. As they had only part of the story, they needed to understand how the story went with the others' translation parts. Even if students did not translate their story well, their summaries were useful in many cases. Students could guess the plot from their reports of translation activities every week, even if their translations were not fully correct. I expected students to guess the important points and the whole story by hearing the other members' summary reports. In this process of reporting on their translations, they could collaborate to understand the whole story every week, and at the same time they found common elements—that is, proper nouns and character names—that should be standardized in their translations.

Third, I used Google Drive's spreadsheet function every week to answer students' questions on their translation. Google Drive is a free cloud storage program with up to 15GB of space, in which people can use various types of interactive documents. It offered an Excel-style sheet as one of the options of Google Drive's cloud service, which could be shared and edited by the participants. I used Excel-type files called "Spreadsheets," for my question-and-answer activity because I could put the whole text of *The Little Brick Church* on the second column of the spreadsheet, students could write their grammatical questions on the left side of the sheet, and I could put the answers to their questions on the third column. All changes were immediately saved on Google Drive. In this case, if students asked translation questions in Japanese they were answered in Japanese.

In many cases, students' translation questions were practical and needed personal direction. Before the introduction of Google Drive, I had used an oral question-and-answer style in class, but it consumed too much time that was not useful to the other class members. In class, translation ability was the main theme of the evaluation, but the students' personal questions on the translation were not properly answered because of the shortage of time. Additionally, sometimes I showed the model translation and then my students copied it. This was not the ideal way for my students to develop their own translation abilities.

After I introduced Google Drive, however, I could manage students' questions in satisfying ways for both me and my students. It took time for me to answer students' questions, but I was able to use the brief moment during the other tasks to answer students' questions on the Internet. I limited students' questions to three a week in order to distribute my answers equally to students. I expected the quality of the translations would improve by using Google Drive's spreadsheet. All the students shared the same spreadsheet, and they were able to check the other students' questions and answers on the translation task. Once this spreadsheet was shared, I found I could more easily communicate with my students.

In Seminar, students cooperated with each other to translate *The Little Brick Church*. Students differed in their translation abilities. In addition to a few good translations, some students' translations were full of errors, and others did crude word-to-word translations. Because this project was a translation of one novel, students compared their translations to find their differences. They realized that their translation styles were so various that they needed some collaborative brushing-up of their translation abilities. My class found there were not only differences in their English abilities but also differences in their styles when translating the same work. They realized that they needed the same tone and same terms when sharing this task.

Revisions of the other students' works were important for them to create a totally collaborative translation, but refining the translation was very difficult to do in the classroom. As a solution to this task, I divided each person's translation text into fourteen parts, one for each of the Seminar's fourteen students, and distributed each two- or three-page translation to the other students, again to improve the translation. This did not promise the completion of the translation, but it still helped improve it. Through this cooperative work, students found they needed to share their task with others.

They also shared their translation drafts on Google Drive's "Documents" to allow them to edit the other members' drafts, but still improving the total quality of the translation was difficult when we translated one work collaboratively. As Tharon W. Howard argues in *Design to Thrive* (2010), "Pulling together activities requiring *collective action* is even more complex and even more difficult to organize than activities involving cooperation."[30] To better integrate the individual translations, Google Drive's interactive function was useful because class members could frequently check others' drafts to edit them. They could add comments or edit errors on the Google Drive's document using the "comment" function. After the other members' comments, the original translator could change the text again. Sometimes the comments were still incorrect, but these different interpretations gave the students the chance to reflect on and rewrite their translated sentences. Google Drive was helpful for students to collaborate on their translation work.

Students' translation abilities improved with these collaborative translations. This exchanging process formed a shared work, and in many cases the collaborative work was better than the original. Students began to understand their writing was not theirs alone, but could be improved with collaboration. After this first task to improve translation, my students and I gathered to discuss the tones of the character's voices. Though the translation activity started as an individual task, the translation was eventually shared and improved by all students. Google Drive helped encourage students' autonomous improvements of their translations through interactive activities on the web. The practical goal of this collaborative translation was to understand the story of *The Little Brick Church*, but this task also helped students understand the effect of collaboration.

Summary of Translation Activities in EALC and Seminar

In two ESL classes, translation activities were conducted to check the effect of Web 2.0 tools (WordPress and Google Drive) on students' translation abilities. In EALC, WordPress was useful to give students the model translation and help them to comfortably compare it with their own translations. In Seminar, Google Drive was used to enhance the interactions between me and my students as well as between the students and their peers. In the latter case, I needed more careful arrangements for students' tasks, and my students were more pressured. However, the students' collaborative activities showed their ability to improve their own translations. In both classes, students translated W.C. Falkner's works by themselves, and they discovered how two Fa(u)lkners' works are connected through social combination.

I used these translation and reading activities to give enough chances for students to understand the contents of their literature and to improve their translations through Web 2.0 tools. After these reading activities, students were ready to express their own opinion in Japanese. At the undergraduate ESL level, if teachers want to use English texts that have not been translated, it is difficult to expect students' autonomous responses and discussions in English immediately after their text readings. Their discussions need Japanese at that level. What I did in these two classes was enhance students' ability to understand the two Fa(u)lkners' works through translations, to discuss these literary works in Japanese in the classroom, and then to write a paragraph on Moodle in English. In these two class activities, WordPress supported students' understanding of the story, and Google Drive encouraged the class members' collaboration.

Evaluation of Students

For the grade calculation in my two classes, I checked students' written paragraphs on Moodle, their weekly translations, and their paper reports on the class texts. My evaluations of these activities included students' Japanese and English abilities. In each session, students translated English into Japanese and submitted their translations in Japanese. Their final reports on the literature were often written in Japanese. Their SCS paragraphs showed their English composition abilities. Sometimes the students' activities revealed their linguistic abilities to be disharmonious. Some students showed keen analyses on the literary texts in their Japanese reports, but their translation exercises and SCS paragraphs showed their English skills were below average. In the other case, I had a student whose English skills were higher than average, but his interpretation of the literary works and his presentations in Japanese were some of the poorest and shortest ones in class. As I wanted to properly judge my students' ability in the translation classes where they used both Japanese and English, I treated the students' English and Japanese activities in an equal percentage and I took an average to give them their grade.

While I was calculating the students' grades, I reflected on the whole-class activities and asked myself whether I realized democratic conditions in these two classes. Teaching democratic relationships among individuals is a crucial theme in my pedagogy. In the following section, I explain why education of the democratic relationship is the ultimate purpose of my American literature courses, and why American literature, especially the literature of the American South, needs this purpose in class.

DEMOCRACY AS A MAIN THEME OF AMERICAN LITERATURE

In today's digital age, ESL teachers need to reconfigure their method to teach American literature in Japan to inspire students' autonomous attitude toward English texts. In general ESL classes in Japan, students are reluctant to say their opinions in class, and feel ashamed to reveal their shortage of knowledge in front of the other students. What they are afraid of is making errors. They become passive in their studies, and they do not ask questions about their readings and interpretations. As Michael Herriman confesses in "Autonomous Learning in EFL in Japan" (2007), most ESL students in Japan do not have autonomy in their English studies. In teaching at a university in Japan for ten years, he confronted teacher-reliant students daily. Most students Herriman encountered had no idea how to take responsibility for their own learning; ESL classes in Japan were mainly instructional and "profoundly a one-way process."[31] This condition is more serious when English classes are taught by Japanese ESL teachers. Since Japanese ESL teachers instruct the target language, English, in Japanese, many ESL students cannot use English autonomously. Teachers use Japanese to explain the meaning of English, and there is little need for the teacher to use English to communicate or discuss with the students. Often English grammar and interpretations of stories are taught in Japanese. In these cases, American literature is studied in teacher-oriented, information-oriented classes. ESL students in Japan expect teachers will give them everything they need to understand about American literature, including grammar, new vocabulary words, and model translations. In this teaching, teachers demand to use canonical texts that will not confuse the students with wrong ideologies. The minor politically, racially, and religiously biased texts tend to be eliminated from teachers' choices.

ESL class should have a democratized atmosphere to study some ideologically biased texts. In the same way as Bakhtin uses dialogue to read Russian literature in a democratic way, today's students need to use democratic dialogue in class to read the literature of the American South from the late nineteenth to the early twentieth century, which are full of discourses that affirm racial segregation and white supremacy. As Wells argues, in many popular Southern works during the several decades following the end of Reconstruction in 1877, "visions of the old plantation became so prevalent."[32] If I want to use these popular Southern fictions in class, I must change my attitude from information-oriented teaching to a communication-oriented approach. Once students build equal relationships with each other in class, any text can be read, studied, and argued. Through temporal realization of their democratic communication, ESL students can discuss any American literary work and understand that individuals' sense of personal democracy is the need to understand the various themes of American literature. In the following argument, I summarize how I taught two Fa(u)lkners' texts in ESL class and what kind of results I achieved.

Teaching Two Fa(u)lkners' Works to ESL Students

I use two terms, monologism and dialogism, to explain the features of both Fa(u)lkners' writing in ESL classes. W.C. Falkner's works have a typical monologic structure, while Faulkner's works are categorized as dialogic. This categorization improves students' understanding of the stylistic development from W.C. Falkner's works to Faulkner's works. After students categorize these works, they can understand how Faulkner criticizes his great-grandfather's writing.

In "Bakhtin and the History of the Language," Tony Crowley writes, "dialogism—monologism are key themes in Bakhtin's work in general."[33] The romantic novel, which shows "the absence of any distance between the author's position and the hero's position" (PDP 84), is a typical example of monologism. In contrast, in the dialogic area, the "orientation of a word among other words (of all kinds and degrees of otherness) creates new and significant artistic potential in discourse, creates the potential for a distinctive art of prose, which has found its fullest and deepest expression in the novel" (DN 275). In the romantic novel the author gives an imperial power to a heroic character; the dialogic novel is an art form which realizes the potential of the democratic words. The tendency of W.C. Falkner's works to monologically simplify the world is criticized in the dialogic structure of Faulkner's works. Only the author's/hero's narrow, white-male-oriented, and self-sufficient voice governs W.C. Falkner's literary world. This worldview may unify his works, but it cannot give any equal rights to dialogic voices.

Faulkner's socially marginalized characters hope to communicate equally with each other, such as Wash Jones in "Wash" or Lucas Beauchamp in *Intruder in the Dust*. They struggle to have common ground with the other characters. In Wash's case, his hope comes to a tragic end, while Lucas in *Intruder in the Dust* temporarily succeeds in having common ground with others. Contrasting them with W.C. Falkner's monologic characters causes two different types of responses in a class. Some students prefer W.C. Falkner's heroic characters in the same way they enjoy the romantic heroes' actions in Hollywood movies or Japanese popular comics. These students prefer the simpler worldviews, the brave actions of heroes and heroines, and the clear conclusions. Interestingly, these students know that Faulkner's works are more highly evaluated. They understand that the monologic worldview in W.C. Falkner's works is criticized and parodied in Faulkner's modern writing; however, their opinions honestly reveal their preference for W.C. Falkner's heroic stories and simple outlines. In contrast, the students who prefer Faulkner's polyphonic novels appreciate Faulkner's critical view on racial segregation and the South's romantic heritage. These differences of the students' preferences are an interesting theme to discuss in our classes.

From Social Combination Activities to SCS Paragraph Writing

I applied the idea of social combination to students' writing activities on the web. In two classes, I expected students to find the way to use the web as a place to create their own opinions by collaborating with their classmates and sharing their writing. Some scholars expect the dialogic relationship between multiple social texts will be realized in today's digital environment. For example, in *Hypertext 3.0* (2006), Landow argues that the dialogic relationship among texts needs "electronic linking, which emphasizes making connections."[34] Or, as Sharon Tettegah suggests in "Computer Mediated Social Justice: A New Model for Educators" (2002), dialogic discourse can be mediated "through human computer interaction."[35] These arguments, however, also need more practical methods in order to be applied to class activities. From my point of view, Web 2.0 has not developed a system to connect students' dialogic activities in class with their writing on the web. Web 2.0 systems share and exchange their texts, photos, and videos, but they do not encourage the users to hybridize their own opinions with others. SCS activity helped students compose their own English paragraph while being aware of the other students' opinions.

SCS activity encouraged students to be the original source for their information. When they created their social combinations in their discussions or compared their ideas on the questions, they found the solutions only came from their communications. They used their brains to create their own ideas or connections. After they developed their own opinions by connecting what they read with what they discussed, their ideas became more rich and persuasive. For example, one of my students, who preferred W.C. Falkner over Faulkner, started to explain his preference for W.C. Falkner's romantic story. In his argument, his writing developed from simply copying the ordinary comments on the literary text to using his own voice to explain his feelings to the others. He compared his preference for W.C. Falkner's work to his fondness for hip-hop, explaining that he likes simple messages, masculine attitudes, and aggressive actions. He recognized this preference by communicating with his peers.

Another student argued that W.C. Falkner's monologic worldview was criticized in Faulkner's works in the same way Hollywood's heroic stories were criticized by her mother, who despised their simple expressions of masculinity. She commented that reading these two stories helped her understand why her mother did not like romantic Hollywood movies and complained that such movies were so-called "American" movies. Connecting her reading experience with her mother's comments on Hollywood movies, she recognized her parent's words as the other's opinion, and understood that the typical Hollywood movie shared the same romanticism and masculinity with nineteenth-century romantic novels.

In my ESL classes, the social combination activity was practically applied to guide ESL students in having equal relationships in their small group works and paragraph writings. When I used this activity for students to discover new concepts, key terms, or perspectives to bridge two different opinions, I heard some students complain they were exhausted because they used their brains too much. Their comments suggested that the social combination activity required the participants to find new ideas without relying on given answers or rules. It seemed that students understood the effects of social combination on practical exercises rather than from lectures.

The students' personally democratic activities aimed to change students' relationships and their ways of thinking in class. If they found their linguistic or physical conditions made it difficult to have immediate equal ground, they should develop that common ground over time. This was one of the most challenging tasks in the social combination activity, and yet I felt it was incredibly important. This feeling came from the regrettable experience I had when I had a blind student in a seminar. When I taught her six years ago, I did not have effective communication activities for students. Because of this, some of the students did not communicate with the blind student while they were taking my seminar. I could have had some supports from SGU for her study, but I only used the translation supports to convert the texts for her into Braille and vice versa. My aim was to conduct my class in the same way I had for years—meaning I acted as if there were no blind student in my class.

I believed that I treated all my students equally, but I soon found I was wrong. Near the end of the semester, I had a party with the students. Almost all the students took part in the party, and became relaxed and enjoyed. I felt I had created a good atmosphere for the seminar members. While I was talking with my students, I casually referred to my glasses. Then my blind student suddenly said to me, "Oh, you're wearing the glasses!" "Yes," I responded. At that moment I understood that she had not known I wore glasses. I let her touch my face, and she recognized my face and glasses with her fingers. While she was putting her fingers on my face and glasses, I realized I had not known her world. Maybe she needed to communicate with the other students more in different ways—in her ways. The other students should know how her worldview is different from theirs. I realized that my methods had not been enough to satisfy her demand to communicate with others.

How could I teach what she wanted? Ironically, for the first time I felt that my blind student and I were standing on the same ground. I did not know her world; she did not know my world. We needed to begin our communication by understanding we did not know each other completely. This mutual unknowing could be overcome by our personal efforts to have the common plane among individuals. But, regrettably, I did not have practical methods then to help students communicate in egalitarian ways. After the party, I changed my methods a little. I believed pair work would create these equal conditions. But I soon found that if I let students choose their own pairs, they created deeply stratified groups in their small class. Many students chose the same partner every time, and so they did not exchange words multiple members of the class. As all students have particular preferences when communicating with other students, free pair work activities tended to create a mutually segregated condition. I felt that we needed some rules or concepts to help students communicate with multiple class members. I also wanted to explain *why* we needed more equal communications with each other in class. This is one of my main motives to create the concept of social combination for my classes.

A personal recognition of equality with others also helps students recognize their differences. I wanted my students to make equal relationships with each other so they could realize their mutual differences. Consequently, when I applied my concept of social combination, I encouraged students to find differences between their opinions. Their new ideas came from efforts to bridge their opinions' differences. To make students communicate with as many students as possible, I limited the time of communication. Rearranging the students' combinations frequently to give them multiple chances to communicate with other class members, I told my students to communicate equally as much as possible. Though I did not have democratic activities when the blind student was in my class, I now compensate by including many social combinations in my classes.

One of the main aims of studying American culture in Japanese ESL classes is to learn how to construct democratic space in personal connections. America is the place where people explore various formations of community to make their conditions better and enhance equality and freedom of speech. If students want to find their own voices in class, they must learn a method to create democratic space for their dialogic discussions. When they exchange their opinions within an egalitarian environment, the differences of their positions appear and they discover new ideas to bridge those differences. Through their struggle to find common ground with others, students understand that the history of American literature is an illustration of that process. Teaching democracy as one of the main themes of American literature gives a clear purpose to students' class activities and inspires students to learn its language, history, and cultural background.

A Definition of Digital Literacy

In *Digital Literacy* (1997), Paul Gilster defines digital literacy as "the ability to understand and use information in multiple formats from a wide range of sources when it is presented via computers."[36] Almost two decades have passed since this definition was written. A new definition of digital literacy is needed to describe today's digital world. In today's Web 2.0 condition, students "write directly online in a blog and get immediate feedback from peers and others who could be anywhere."[37] They enjoy Facebook, Twitter, Instagram, and other local SNSs. However, these digital platforms do not develop students' digital literacy in academic ways. As Brabazon argues, "Searching must not be confused with literacy. Finding content is not necessarily understanding context. Sharing is not a proxy for social change. Liking a link or retweeting 140 characters does not signify social change."[38] The imperative of teachers who teach today's digital literacy in class is to change students' ideas on their communication with the world through digital networks.

Today's interactive, digital condition requires a new definition of people's communication online. I asked students in my two classes to discuss a new definition of digital literacy through their social combination activities. Students discussed and created their own definitions. Some students concluded that today's digital literacy is the ability to collect true data and information. Many students agreed. This conclusion adds a new element to the definition of Paul Gilster. However, this answer also suggests a more negative tendency of students: no students felt digital literacy is a tool to create their own opinions with others. Students do not focus on the social aspect of the digital network, which is an effective tool to create new ideas from interactions with other people. In my class, I needed to use some practical activities to help my students realize how digital literacy is effectively used to create and publish new ideas.

I argue digital literacy in Web 2.0 is the ability to understand and create digital information in interactive formats on the web. As Solomon and Schrum write in *Web 2.0* (2007), the guideline for technology-based student activities, the National Educational Technology Standards for students, insists on creativity and innovation as the first category of student achievement; its purpose is "students demonstrate creative thinking, construct knowledge, and develop innovative products and processes using technology"; and the first and foremost ability students should develop is to "apply existing knowledge to generate new ideas, produces, or processes."[39] In digital literacy, understanding as well as creating digital information is important. This creation includes the ideas that come from our interactive communications on the web. Today people are tweeting, writing on Facebook, commenting on blogs, and chatting with friends online. Digital literacy includes these everyday communications on the web, but also includes people's higher ability to establish their own arguments by integrating others' opinions. Students' discussion on the definition of digital literacy suggested that this aspect was not well recognized among students, even if they used Web 2.0 every day.

Following the new definition of digital literacy, my pedagogy in ESL class intends to help students be aware of the creative aspect of the web by integrating multiple viewpoints. One's opinion should contain several viewpoints or at least be aware of other points of view. It is still necessary to teach students how to invent their combined ideas using digital equipment: one of the best ways to invent new ideas is to communicate with others. Students can gain new ideas by exchanging their opinions with others on the web in the same way they do in their everyday life. My class activity is a prototype for helping students prepare for this intersubjective condition that should be realized on the web.

SUGGESTIONS FOR FUTURE CLASS ACTIVITIES

In my two classes, I used three methods—social combination activities, SCS paragraph writings, and digital translation activities—to explore how to increase ESL students' understanding of multiple literary works. These three methods are applicable to various literary classes—I hope teachers find the English-Japanese translation activities that use WordPress and Google Drive can be applicable to various two-language translation activities. The social combination activities in class and paragraph writings on Moodle are effective in ESL classes with multicultural students. Since one of the main purposes of social combination activities is to increase students' recognition of different viewpoints, teachers can apply this method to make students communicate with others who may have different backgrounds. The SCS paragraphs also help students recognize what kind of dialogue they had according to their texts in class.

In ESL classes with multicultural students, comparing two different authors' literary texts increases the students' comprehension of the literature. I believe comparing two literary works that come from the same region can give great hints on the historical and geographical background of that region, but it is also important to choose literary works that have some thematic relationship when the teacher want an educational effects from such comparison. For example, comparing Elizabeth Stoddard's *The Morgesons* (1862) with Grace Metalious's *Peyton Place* (1956) is an interesting comparison of the tradition of romance in New England. And when the teacher wants to explore transatlantic influences on Stoddard's work, Charlotte Brontë's *Jane Eyre* (1847) gives a good early effort. This helps ESL students understand how the British novel influences the American one in the nineteenth century. If a teacher wants to pursuit the development of greener theme rather than regional connection, Henry David Thoreau's *Walden* (1854) and John Muir's *My First Summer in the Sierra* (1911) help explore the theme of ecocriticism.

The Morgesons, *Peyton Place*, and *My First Summer in the Sierra* have not been translated into Japanese. If these works are to be read by Japanese ESL students, translation activities using WordPress and Google Drive can help guide students' understanding of these texts. If it is impossible to translate the whole text, translating extracts and comparing them with the other text's excerpts can develop students' translation abilities and understanding of the stories' some important parts. Teachers can use WordPress, Moodle, or any other free blog tool to show his own model translation to students. In this case, the class activities tend to be teacher-oriented, and the teacher can request a final report on students' comparison of two texts. The grades are scored based on the students' usual activities and final papers.

If teachers want to encourage their students' autonomous translations of *The Morgesons*, *Peyton Place*, or *My First Summer in the Sierra*, they can use Google Drive to have interactive and cooperative translation activities with students. However, it is highly possible that students' translations do not give a total comprehension of the text. In this case, students' translation experiences should be combined with social combination activities and SCS paragraph writings. Using these methods, teachers can check on students' discussions and their understanding of the texts. Of course, if some students want to challenge advanced studies or total translations of *The Morgesons*, *Peyton Place*, or *My First Summer in the Sierra*, teachers can encourage these challenges.

If teachers have multicultural students in their ESL classes, translation activities are not the proper method to instruct students—social combination activities and SCS paragraph writings are more effective. Translation activities using WordPress and Google Drive should be limited to classes whose students share the same first language. However, social combination activities and SCS paragraph writings are widely applicable and encourage class discussions, reflection, and original composition.

CONCLUSION

I applied the concept of social combination to my literary instruction. The idea to create common ground among students was practically exercised in the following two methods: social combination activities and SCS paragraph writings on Moodle. These two methods developed my students' ability to have democratic arguments with the other class members, compose ideas by including other voices in a paragraph, and improve their translations. This chapter argues that social combination is a basic concept for ESL students to share their opinions with other class members. Without this dialogue, students do not realize how their opinions differ from others. I wanted my students to develop their own ideas, their own writings, from these discussions based on their differences, and find others' voices a great resource for their writing.

The purpose of these social combination activities was to develop students' egalitarian communications with others in practical steps. In the beginning of the course, students might have been nervous to communicate with other unknown students. Social combination activities reduced students' psychological pressure regarding enforced communications in class. It is natural that a person feels apprehension in communicating with an unknown partner. Following social combination's rules, before their discussions began, I advised students that there was no need for them to be friendly with each other. And these social combination activities lasted only five or ten minutes. They were asked to create some responses to particular questions in this limited time, and there was no need for them to communicate about anything else. Friendship and fraternity were positive results of these class activities, but they were not their purpose. To communicate equally and gain new ideas were what social combination activities demanded. I gave this advice to the students before their collaborations and found good results in the class atmosphere. Maybe this advice is more effective to make students' multicultural conditions communicative in ESL international classes.

The results of SCS writings showed students could develop their own voices when they reflected on their partner's opinions in their paragraphs. Having a social combination activity in class and an individual reflection activity on Moodle provided a good balance to students' studies. My goal was to bridge students' discussions in class and their writing activities on the web. Although some students skipped a few proper procedures on Moodle, I was satisfied with my results. The main purpose of SCS writings was to help students understand that writing a paragraph is the result of encounters with other voices.

In these two classes, the translation activities of W.C. Falkner's works were conducted with the support of two digital platforms: WordPress and

Google Drive. These two platforms had different effects on my activities. In EALC, students summed up ten pages of *The Spanish Heroine* every week, and they checked their errors and story details in a Japanese translation that I updated weekly on WordPress. All of EALC students liked this system. Students could check the story of *The Spanish Heroine* any time they needed. When they wrote their reports, the students who chose *The Spanish Heroine* for their report read the Japanese version on WordPress and quoted the translation as well as the original English. Some students loved *The Spanish Heroine*. Almost all of my students understood what kind of romantic story W.C. Falkner created in his imagination. They compared *The Spanish Heroine* with Faulkner's short story, "Wash," to understand how Faulkner's characters were influenced by the nineteenth-century romantic image of the brave Southern hero.

In Seminar, my students used Google Drive to support their translation. In contrast to EALC, I did not model the translation of *The Little Brick Church*. Instead, I used Google Drive to answer students' questions on their translation activity of *The Little Brick Church*, and students were expected to complete their translation by cooperating with the other class members. After they completed their first draft, they shared that draft on Google Drive and other students added comments. Their cooperative activities necessitated carefully planned tasks for the students and put a large amount of pressure on them to complete these tasks, but the translation activity on Google Drive increased the quality of their translation. At the end of class, students gave positive feedback on their digital translation activities.

Students in both classes responded positively to my pedagogical activities. EALC students enjoyed reading W.C. Falkner's *The Spanish Heroine* and Faulkner's "Wash." Students communicated through their social combination activities and composed their opinions in their SCS paragraph writings. In Seminar, students completed a first draft of the translation of *The Little Brick Church* and then compared it with *Intruder in the Dust*. Reading multiple authors' texts with digital supports increased students' understanding of a particular region's literature and its historical and cultural context. Social combination activities, SCS writings, and digital translation activities helped me develop students' ideas through their dialogues in class and their activities on the web.

NOTES

1. Todorov, *Mikhail Bakhtin*, 30.
2. Kiran Kaur Dhillon and Surita Mogan, "Language-Based Approaches to Understanding Literature: A Creative Activity Module," *English Teacher* 43, no. 2

(2014): 63; Asha Rai, "Use of Literature in Teaching English," *International Journal of Educational Research and Technology* 3, no. 3 (2012): 71; Paula Wolfe, "'The Owl Cried': Reading Abstract Literary Concepts with Adolescent ESL Students," *Journal of Adolescent & Adult Literacy* 47, no. 5 (2004): 402; Matthew Hodge, "'Bicycles, Muscles, Cigarettes,' and ESL: Raymond Carver in the Classroom Matthew Hodge, Central Piedmont Community College," *Teaching American Literature* 7, nos. 1–2 (2014): 44.

3. Duclos, *Son of Sorrow*, 9.

4. Linda Gajdusek, "Toward Wider Use of Literature in ESL: Why and How," *TESOL Quarterly* 22, no. 2 (1988): 227; Moussa Traore and Lydia Kyei-Blankson, "Using Literature and Multiple Technologies in ESL Instruction," *Journal of Language Teaching and Research* 2, no. 3 (2011): 562; Gülşah Tikiz and Feryal Çubukçu, "Symbiotic Interface between Literature, Culture and Language Learning," *Journal of International Social Research* 6, no. 26 (2013): 134; Hodge, "'Bicycles, Muscles, Cigarettes,' and ESL," 44.

5. Gajdusek, "Toward Wider Use of Literature in ESL," 234.

6. Hodge, "'Bicycles, Muscles, Cigarettes,' and ESL," 45.

7. Tikiz, "Symbiotic Interface between Literature, Culture and Language Learning," 134.

8. The movie version of *Intruder in the Dust* is a high-quality adaptation of the novel, and it shows the novel's theme in a more concise way. As Li argues, the novel and its film version have necessary differences, but both are mainly concerned with "the emergent social and personal consciousness of Chick . . . who learns to negotiate and locate his own sense of self within the culturally ingrained conventions and expectations of his Southern community. See Li, "*Intruder in the Dust* from Novel to Movie," 105.

9. Cash, *The Mind of the South*, 44, 379.

10. Robert Petrone and Robert Gibney, "The Power to Speak and Listen: Democratic Pedagogies for American Literature Classrooms," *English Journal* 94, no. 5 (2005): 39.

11. Stephen Sharp, *A New Look at the Interactive Writing Classroom: Methods, Strategies, and Activities to Engage Students* (Lanham: Rowan and Littlefield Education, 2011), 118, 123.

12. Jessica Labbé, "'Death by Misadventure': Teaching Transgression in/through Larsen's *Passing*," *College Literature* 37, no. 4 (2010): 120.

13. Sharp, *A New Look at the Interactive Writing Classroom*, 124.

14. Alexis de Tocqueville, *Democracy in America*, trans. Arthur Goldhammer (1835, 1840; New York: Library of America, 2004), 5.

15. Petrone, "The Power to Speak and Listen," 39.

16. Venessa Paech, "The Inevitable Exile: A Missing Link in Online Community Discourse," in *Digital Dialogues*, 12.

17. Hannah Arendt, *The Human Condition* (1958; Chicago: University of Chicago Press, 1998), 41.

18. Gergen, *The Saturated Self*, 107.

19. Thomas A. Upton, "First and Second language Use in Reading Comprehension Strategies of Japanese ESL Students," *TESL-EJ* 3, no. 1 (1997): 19.

20. Goh Hock Seng and Fatimah Hashim, "Use of L1 in L2 Reading Comprehension among Tertiary ESL Learners," *Reading in a Foreign Language* 18, no.1 (2006): 30.

21. Suppasetseree and Dennis tell that most teachers used Moodle's "*uploading and sharing documents feature in order to assign students' homework and encouraged students to download learning material outside the class to practice their English learning skills*." Additionally, as Miyazoe argues, students were able post and restore their "blogs and essays in English on Moodle." Moodle has not been normally used to develop students' dialogical activities and writing. See Suksan Suppasetseree and Nutprapha K. Dennis, "The Use of Moodle for Teaching and Learning English at Tertiary Level in Thailand," *International Journal of the Humanities* 8, no. 6 (2010): 29; Terumi Miyazoe, et al., "'To-do-or-not-to-do-dilemma' Online: Blog and Essay Writing Visualizations in a Blended-Format University English Course," *e-FLT: Electronic Journal of Foreign Language Teaching* 11 (2014): 165.

22. For more details, check https://moodle.com/cloud/ (Look at Appendix to see how to create the SCS system on free MoodleCloud).

23. Hirschkop, *Mikhail Bakhtin*, 86.

24. Saziye Yaman, "Technology Supported Learning Platform: Moodle Integrated Academic Course," *Turkish Online Journal of Distance Education* 11, no. 2 (2010): 148.

25. See Appendix to check how to create SCS system using Moodle's free hosting service for educators; or visit https://socialcombination.wordpress.com/.

26. Barton, *Language Online*, 159.

27. Judith Oster, "Seeing with Different Eyes: Another View of Literature in the ESL Class," *TESOL Quarterly* 23, no. 1 (1989): 89.

28. Sharp, *A New Look at the Interactive Writing Classroom*, 80.

29. Catherine G. Van Beuningen et al., "Evidence on the Effectiveness of Comprehensive Error Correction in Second Language Writing," *Language Learning* 62, no. 1 (2012): 2; Icy Lee, "Working Smarter, Not Working Harder: Revisiting Teacher Feedback in the L2 Writing Classroom," *Canadian Modern Language Review* 67, no. 3 (2011): 378; Abdulmoneim Mahmoud, "Strategy-Based Peer Assistance in EFL Writing: An Alternative to Traditional Peer Correction," *Theory and Practice in Language Studies* 2, no.10 (2012): 1987.

30. Tharon W. Howard, *Design to Thrive: Creating Social Networks and Online Communities that Last* (Amsterdam: Elsevier, 2010), 19.

31. Michael Herriman, "Autonomous Learning in EFL in Japan," *International Journal of Learning* 14, no. 5 (2007): 142.

32. Wells, *Romances,* 2.

33. Tony Crowley, "Bakhtin and the History of the Language," in *Bakhtin and Cultural Theory*, ed. Ken Hirschkop and David Shepherd (Manchester: Manchester University Press, 2001), 180.

34. George P. Landow, *Hypertext 3.0: Critical Theory and New Media in an Era of Globalization.* (Baltimore: Johns Hopkins University Press, 2006), 113.

35. Sharon Tettegah, "Computer Mediated Social Justice: A New Model for Educators," *TechTrends* 46, no. 5 (2002): 30.
36. Gilster, *Digital Literacy*, 1.
37. Solomon, *Web 2.0*, 2.
38. Brabazon, "Conclusion," 256.
39. Solomon, *Web 2.0*, 251.

Conclusion

Personal Democracy in Social Combination

This study uses the term social combination to explore the ways to study and teach two Fa(u)lkners' works with ESL students in today's digital conditions. The idea of social combination is inspired from Bakhtin's dialogic concept in which language is not a unitary whole but contains contradictive worldviews of various people. Though Bakhtin presupposes that multiple voices can be juxtaposed on a common plane in a novel as if they were symphonized in the orchestra hall, I reinterpret that his dialogic common plane can only be realized in personal efforts to have equal relationships between two or three individuals because of the world's fragmented condition after September 11, 2001. Social combination means individuals' personal effort to have democratic relationships for a while, to suggest people cannot eradicate their stratified conditions but still have ability to have temporal common grounds with others. It is hard for each Fa(u)lkners' character to conceive of the long-lasting realization of personal democracy, but their temporary social combination suggests which direction their dialogical relationships aim at. In this book, the concept of social combination connects two Fa(u)lkners' works with ESL students' process to create dialogic spaces with others both in class and on the digital network.

The idea of social combination helps us understand two Fa(u)lkners' works and digital literacy in a continuous historical process. In W.C. Falkner's romantic works, a pattern is fixed: some characters temporarily develop social combinations in the middle of the story, but by the end all of their relationships are absorbed into the formation of a narrow community under white male dominance. W.C. Falkner's romantic works describe America as a country of patriarchal democracy, which is supported by the ideological belief that the white elites, who are the rightful inheritors of Anglo-Saxon culture, are the ideal dominators of the land, the nation, and even the world.

This belief approves the white men's violent wills to protect and dominate the other socially marginalized members in their narrow communities. In contrast to W.C. Falkner's works, his great-grandson Faulkner sees the history of the American South as a series of failures of people's ambitious efforts to create social organizations. His modern literary works illustrate that people's desire to make better social formations may end in failure, yet people still endure. Faulkner's historical perspective and his positive evaluation of the possibilities of personal relationships resist the monotonous illusion of romantic unity in W.C. Falkner's works. Because of this historical development, two Fa(u)lkners' works are appropriate texts to study for ESL students whose communication network has been digitally expanded. Today's ESL students are confronting two serious challenges. On the one hand, students have to explore equal connections and common rules in their globally expanded networks. On the other hand, they demand to retain their sense of diversity based on their local sense of belongings. There are no ideal answers to these two challenges, but the historical development of two Fa(u)lkners' works encourage students to explore the new arrangements of their community retaining their personal democracy and avoiding the danger of falling into the desire to build up self-centered hierarchies in their small communities or to have eternal agreements with others. Instead of complete unity, Faulkner offers social combination as an important process of our communal formations in *Intruder in the Dust* and his later works. Studying two Fa(u)lkners' works with the concept of social combination helps today's ESL students reassess their globally expanded communications with a proper historical perspective.

Faulkner criticizes W.C. Falkner's romantic imagination because it aims to form a narrow community under the white hero's violent dominance. To find this relationship, chapter 1 explained how Faulkner's works re-accentuate W.C. Falkner's motifs. In Faulkner's early and middle works, the white male subjects are caught in the gravity of the heroic imagination that W.C. Falkner's works typically represent. To criticize his great-grandfather's imagination, Faulkner chronicles his modern characters' tragic failures to compose their ideally unified community in the American South. Their goal to create a hierarchical order under the dominance of a white male fails because the other socially marginalized members resist such arrogant control. Depicting this tragic pattern as Yoknapatawpha's history, Faulkner suggests that people's struggle to form new societies based on their personal designs creates the history of the American South, even if those attempts end in failure.

It is effective to juxtapose W.C. Falkner and Faulkner in order to understand how their works evolved from monologism to dialogism. In Bakhtin's argument, monologism typically appears in the romantic novel and reflects the author's self-oriented perspective—the hero of the story is used as a tool to represent the author's idea. In contrast, dialogism is where multiple characters confront each other to reveal their internal and external differences. Reconnection of these two authors' works using Bakhtin's concepts reveals how Faulkner's novels undermine W.C. Falkner's worldview. The authoritative power of W.C. Falkner comes from his monologism, which amalgamates the white male protagonists' violent desire for richness, segregation against the nonwhite races, and imperial invasion with his age's romantic illusion. This vision seems to be successfully realized in both W.C. Falkner's life and works, even though today's biographical studies betray his criminal acts, interracial relationships, and economical, political feuds. Southern romantic worldview supports and reproduces the white people's violent passions, supremacies, and excessive patriarchies. This literary tradition in the antebellum South survives after the Civil War to support the impracticable dream of W.C. Falkner. Faulkner uses his great-grandfather's romantic imagination and patterns as the central cause of the moralistic corruptions of his modern characters.

In chapter 2, *Intruder in the Dust* was read as an introduction to Faulkner's works. In his early and middle works, Faulkner mainly details the white male characters' obsessions in order to criticize their self-content behaviors and worldviews; however, in the initial novel of his later works, *Intruder in the Dust*, Faulkner writes characters that attain positive results in their actions and social formations of their local community. Chick fights against the white community's violent desire for lynching and criticizes this social norm in his dialogue with Lucas, Aleck Sander, and Miss Habersham. Chick's desire to reexamine the community from the other points of view contrasts with the white male protagonists' demand in Faulkner's early and middle works. His dialogues with the socially marginalized characters show their will to overcome the Southern monolithic discourse, which prevents personal communication beyond the borders of gender, race, and class. Their detective process is also a metaphor for Chick's discovery of intersubjective relationships among socially stratified people. Chick, a sixteen-year-old white boy, rejects to be a member of the monolithic society of the white elites, but instead temporarily creates an alternative social form with marginalized people to give another flexible vision on their society.

This novel illustrates that characters' dialogical relationships with others can examine the racial problem of the American South and resist the white males' monologic masculinity. When Chick makes social combinations with other socially marginalized characters, he discovers that the voice of each participant appears as a mutual phenomenon that cannot be completely possessed by one individual. Socially marginalized people express their hopes for democracy by relying on their interpersonal contexts. In Lucas's case, his plea of not guilty in his murder case comes from his silenced suggestions and personal connections with Chick. Chick, who has already experienced the limitations of white supremacy and masculine individualism, perceives Lucas's demand for life as a chance to form a new intersubjective recognition with others. He talks with Aleck Sander and Miss Habersham to confirm the marginalized people's will to resist the masculine dominance of the white male adults. Chick's dialogues and actions with these socially marginalized characters betray that the white males, who are obsessed with dominance in their narrow society, have not formed the personally democratic space to hear the socially marginalized members' complaints.

Chick's action anticipated the historical theme of his region's political struggles. The discourse of the civil rights movements in the middle of the twentieth century aimed to connect personal solidarities with democracy. Martin Luther King Jr. demonstrated that egalitarian relationships in the American South were preserved in people's temporal, personal connections under the pressure of the racial segregation. The rhetoric of King's speech inspired the audiences' immediate and temporal actions to realize the equality in their personal relationships. In his words and demonstrations, King encouraged people's social combinations as important individual actions that coincided with the political changes in the world. People's efforts during the civil rights movement suggested the radical possibility of the personal democracy in the American South.

Faulkner's own attitude toward racial segregation in his community was conservative, but it is applicable to today's globalized condition. Faulkner himself assumes that people still have historical and cultural limitation on their democratic attitude, but individuals' search for common ground in their dialogic spaces help us have a chance to see the world in alternative ways. Historically limited individuals crossover the borders of race, class, and gender to create temporal common grounds, on which people can search for their better lives. Individuals have social combinations to examine the world from other points of view, even if they cannot completely change or alter their own cultures and customs. *Intruder in the Dust* suggests that if people stop their efforts to form personal democracy in their community, individual actions waste their meanings and lose their connections with the world. Under the pressure of the dominant social power that deepens the community's stratified condition, Chick enters into the "vast teeming anonymous solidarity of the world" (ID 203), through his feeling that he shares the same sensations (such as eating) with others. Because of his pursuit of personal connections through common sensations, Chick can be interpreted as a contemporary figure. Today's people also form personal relationships with others through their shared sensations on the web, while their digital network also boosts the world's deeply stratified condition.

Chapter 3 reported two class samples where an ESL teacher connects the texts of W.C. Falkner and Faulkner using today's digital environment in class activities. This report explored methods of teaching American literature, especially the literature of the American South, in Japanese undergraduate courses. Both Fa(u)lkners' works were used to help students understand the literature of the American South as a process to explore ideal social forms. The teacher instructed students to create social combinations in class to resist the authoritative pressure of English monologic learning and see Faulkner's works as a parody of the romantic worldview of W.C. Falkner. This pedagogy aimed to increase students' interest in the literature of the American South and connect it to their own life and social formations. Starting with the relationship between W.C. Falkner and his great-grandson, students could expand their understanding of the development of literature in the American South. W.C. Falkner expressed the romantic tendency of Southern culture in the nineteenth century. Each of his stories was a simple romance between a hero and a heroine, who achieved the typical white male dominance of the narrow community. W.C. Falkner's romantic pattern became the premise of Faulkner's modern novels, and studying W.C. Falkner's works helped readers understand Faulkner's works.

Additionally, class activity of translation of American literature aimed to develop the students' translation skills. I coupled one of these minor literary works with a major literary work, which helped students develop their translation skills since the minor one of the works was not available in Japanese. However, this pedagogy sometimes caused ideological problems. Teaching an unfamiliar, minor text was an exciting experience for both the teacher and students, but these texts frequently had political, racial, or religious problems for today's readers and required the teacher's careful attention. For example, the Southern popular novelists from the end of the nineteenth century to the beginning of the twentieth century included Thomas Dixon Jr. and Thomas Nelson Page as their representative figure and W.C. Falkner as one of the minor ones. Their works were underestimated in today's literary history because of their support for racial segregation, white supremacy, and patriarchal discourse. If the teacher wanted to use one of such texts for students' translation activities, the text should be offered to ESL students with caution. The teacher should encourage students to examine the text from various points of view, and then inform that studying the text in today's classroom helped students understand how the dominant, popular discourse of the American South enforced the patriarchal, white-male-oriented power structures and how they shared their ideological positions with broad audience before the appearance of Faulkner's literature. Social combination activities and SCS paragraph writings helped keep students' plural visions on the text and democratic condition in class.

I used social combination activities and SCS paragraph writings to guide the students' discussions to compare and analyze the different authors' works. ESL students—who formed temporary, equal relationships in class through their social combination activities—discussed how two authors' works could be compared or contrasted. Students wrote their paragraphs on SCS on Moodle to reflect their discussions. Results of these SCS paragraphs showed students developed their abilities to reflect the other class members' voices in their English. Social combination activities in class and SCS paragraph writings were prototypes of class exercises that developed students' dialogic activities with others to understand the literature of the American South. Additionally, translation activities to help ESL students understand both Fa(u)lkners' works were supported by WordPress and Google Drive.

The idea of social combination is a way to teach students how to make intersubjective, democratic relationships in class or on the digital network. Having brief democratic relationships in small groups with classmates helps ESL students understand how to create their own dialogic space with others. ESL students also need to develop their ability to write English dialogically on the net because individuals still have difficulties to create their dialogical relationships on digital platforms. As Paech argues, "Social networks such as Facebook or Twitter may host a collection of smaller communities within their walls, but the platforms themselves are not a community."[1] There should be a new communicative formation in our world to sustain our temporal, democratic relationship to make our dialogic words resist various monologic perspectives and imperial impulse, even on the digital network. Whether their activities are held in the classroom or on the web, social combination activity helps ESL students build their interactive relationships and dialogical opinions by including other students' points of view. By using social combination as a basic concept to examine personal relationships, ESL students can learn how to critically study literature along with digital literacy in democratic condition.

NOTE

1. Paech, "The Inevitable Exile: A Missing Link in Online Community Discourse," 13.

Appendix

How to Create the Social Combination Sheet (SCS) System on MoodleCloud Course

1. Visit Moodle for Free Site (https://moodle.com/cloud/free/).
2. Create your account for Moodle for free. It is free for single classes or just trying Moodle out.
3. Log in. Visit the Introduction to Moodle course to create your course.
4. To create the SCS system, click "Turn editing on."
5. Click "Add an activity or resource."
6. Choose "Database" and click "Add."
7. Fill in "SCS TEST" on the "Name" area. Click "Save and return to course."
8. Click "SCS TEST." Then Click "Fields." Click "Create a new field" button and choose "Text area."
9. Fill in "Element A" in Field name area. Change "Width" into 40 columns and "Height" into 20 rows. Click "Add."
10. Click "Create a new field" button and choose "Text area," fill in "Element B" in Field name area. Change "Width" into 40 columns and "Height" into 20 rows. Click "Add."
11. Click "Create a new field" button and choose "Text area," fill in "Your Paragraph" in Field name area. Change "Width" into 40 columns and "Height" into 20 rows. Click "Add." Click "Save."
12. Click "Templates," and delete all following elements in the "Single template" area (see table A.1)

Table A.1 Elements to delete in the "Single template" area

Element A:	[[Element A]]
Element B:	[[Element B]]
Your Paragraph:	[[Your Paragraph]]
##edit## ##delete## ##approve## ##disapprove## ##export##	

13. Click “Show more buttons.” Click “HTML” button. Paste the following HTML text on “Single template.”

```
<table style=“width: 800px; height: 200px;” border=“0”>
<tbody>
<tr>
<td style=“text-align: center; width: 250px; height: 20px;”><strong>Element @A</strong></td>
<td> </td>
<td style=“text-align: center; width: 250px; height: 20px;”><strong>Your Paragraph</strong></td>
<td> </td>
<td style=“text-align: center; width: 250px; height: 20px;”><strong>Element B</strong></td>
</tr>
<tr>
<td style=“border-color: #000000; border-style: solid; border-width: 1px;”>[[Element A]]</td>
<td> </td>
<td style=“border-color: #000000; border-style: solid; border-width: 1px;”>[[Your Paragraph]]</td>
<td> </td>
<td style=“border-color: #000000; border-style: solid; border-width: 1px;”>[[Element B]]</td>
</tr>
</tbody>
</table>
Timetable: ##timeadded##/ TimeModified: ##timemodified## / User: ##user##</p>
<strong>Edit</strong>: ##edit##  Delete: ##delete##</p>
```

14. Click “Save template” button.
15. Click “List template” button, and delete all following elements in the “Repeated entry” area (see table A.2)
##delcheck##

Table A.2 Elements to delete in the “Repeated entry” area

Element A:	[[Element A]]
Element B:	[[Element B]]
Your Paragraph:	[[Your Paragraph]]
##edit## ##more## ##delete## ##approve## ##disapprove## ##export##	

16. Click “Show more buttons.” Click “HTML” button. Paste the following HTML text on “Repeated entry.”

```
<table style=“width: 800px; height: 200px;” border=“0”>
<tbody>
<tr>
<td style=“text-align: center; width: 250px; height: 20px;”><strong>Element A</strong></td>
<td> </td>
<td style=“text-align: center; width: 250px; height: 20px;”><strong>Your Paragraph</strong></td>
<td> </td>
<td style=“text-align: center; width: 250px; height: 20px;”><strong>Element B</strong></td>
</tr>
<tr>
<td style=“border-color: #000000; border-style: solid; border-width: 1px;”>[[Element A]]</td>
<td> </td>
<td style=“border-color: #000000; border-style: solid; border-width: 1px;”>[[Your Paragraph]]</td>
<td> </td>
<td style=“border-color: #000000; border-style: solid; border-width: 1px;”>[[Element B]]</td>
</tr>
</tbody>
</table>
Timetable: ##timeadded##/ TimeModified: ##timemodified##/ User: ##user##</p>
<strong>Edit</strong>: ##edit##  Delete: ##delete##</p>
```

17. Click “Save template” button.
18. Click “Add entry, “ and you can write your “Element A,” “Element B,” and “Your Paragraph” in each column.
19. Click “Save and view” button. Then you can see how the SCS system work on your Moodle.
20. Once you create “SCS TEST” sheet, you can “duplicate” this sheet.

*For more information, visit < https://socialcombination.wordpress.com/ >.

Bibliography

Aiken, Charles S. *William Faulkner and the Southern Landscape*. Athens: University of Georgia Press, 2009.

Anderson, Benedict. *Imagined Communities: Reflections on the Origin and Spread of Nationalism*. 1983. London: Verso, 1991.

Anshen, David. "Faulkner's 'Curious Folk': Estrangement as Critique of Capitalism." *Mississippi Quarterly* 61, no. 3 (2008): 477–95.

Arendt, Hannah. *The Human Condition. 1958*. Chicago: University of Chicago Press, 1998.

Augé, Marc. *Non-Places: Introduction to an Anthropology of Supermodernity*. 1992. Translated by John Howe. London: Verso, 2000.

Ayers, Edward L. *The Promise of the New South: Life after Reconstruction*. Oxford: Oxford University Press, 2007.

Baker, Houston A., Jr. *Turning South Again: Re-thinking Modernism/Re-reading Booker T.* Durham: Duke University Press, 2001.

Bakhtin, Mikhail M. "Discourse in the Novel." 1975. In *The Dialogic Imagination*, translated by Caryl Emerson and Michael Holquist, 259–422. Austin: University of Texas Press, 2011.

———. "From Notes Made in 1970–71." 1979. In *Speech Genres and Other Late Essays*, translated by Vern W. McGee, 132–58. Austin: University of Texas Press, 2013.

———. *Problems of Dostoevsky's Poetics*. 1963. Translated by Caryl Emerson. Minneapolis: University of Minnesota Press, 1994.

Barton, David, and Carmen Lee. *Language Online: Investigating Digital Texts and Practices*. London: Routledge, 2013.

Beuningen, Catherine G. Van, Nivja H. De Jong, and Folkert Kuiken. "Evidence on the Effectiveness of Comprehensive Error Correction in Second Language Writing." *Language Learning* 62, no. 1 (2012): 1–41. http://dx.doi.org/10.1111/j.1467–9922.2011.00674.x.

Blassingame, John W. *The Slave Community: Plantation Life in the Antebellum South*. New York: Oxford University Press, 1979.
Blotner, Joseph. *Faulkner: A Biography*. vol. 2. New York: Random House, 1974.
———. *Faulkner: A Biography*. One-Volume edition. New York: Vintage Books, 1991.
Bondurant, Alexander L. "William C. Falkner, Novelist." In vol. 3 of *Publications of the Mississippi Historical Society*, edited by Franklin L. Riley, 113–25. Oxford, MS: Mississippi Historical Society, 1900.
Boyne, Roy. *Foucault and Derrida: The Other Side of Reason*. 1990. London: Routledge, 2001.
Brabazon, Tara. "Conclusion: White Men Rule?" In Brabazon, *Digital Dialogues*, 253–57.
———, ed. *Digital Dialogues and Community 2.0: After Avatars, Trolls and Puppets*. Oxford: Chandos Publishing, 2012.
———. "Introduction: New Imaginings." In Brabazon, *Digital Dialogues*, 1–8.
Brontë, Charlotte. *Jane Eyre*. 1847. London: Penguin, 2006.
Brooks, Cleanth. *William Faulkner: The Yoknapatawpha Country*. 1963. Baton Rouge: Louisiana State University Press, 1990.
Brooks, Van Wyck. *The Times of Melville and Whitman*. Scranton: E. P. Dutton, 1947.
Calhoun, Frances Boyd. *Miss Minerva and William Green Hill*. 1909. Chicago: Reilly and Britton, 1914. https://archive.org/details/missminervaandwi00calhiala.
Cantwell, Robert. Introduction to *The White Rose of Memphis*, by William C. Falkner, v–xxvii. New York: Coley Taylor, 1953.
———. "The Faulkners: Recollections of a Gifted Family." In *William Faulkner: Three Decades of Criticism*, edited by Frederick J. Hoffman and Olga W. Vickery, 51–66. 1960. New York: Harcourt, 1963.
Capitani, Diane N. *Truthful Pictures: Slavery Ordained by God in the Domestic Sentimental Novel of the Nineteenth-Century South*. Lanham: Lexington Books, 2009.
Carpenter, David. "Faulkner's *Intruder in the Dust*: Apology for South's Jimcrow." Review of *Intruder in the Dust*, by William Faulkner. *Daily Worker* 6 (October 1948): 13.
Cash, W.J. *The Mind of the South*. 1941. New York: Vintage Books, 1991.
Cervantes, Miguel de. *The Ingenious Hidalgo Don Quixote de la Mancha*. Translated by John Rutherford. 1605. New York: Penguin, 2003.
Chopin, Kate. *The Awakening*. 1899. New York: W.W. Norton, 1994.
Cook, Sylvia Jenkins. *From Tobacco Road to Route 66: The Southern Poor White in Fiction*. Chapel Hill: University of North Carolina Press, 1976.
———. "Reading Clothes: Literary Dress in William Faulkner and Erskine Caldwell." *Southern Literary Journal* 46, no. 1 (2013): 1–18.
Cowley, Malcolm. Introduction to *The Portable Faulkner*, vii-xxxiii. Edited by Malcolm Cowley. New York: Viking Press, 1974.
———, ed. *The Portable Faulkner*. 1946. New York: Viking Press, 1974.
Crowley, Tony. "Bakhtin and the History of the Language." In *Bakhtin and Cultural Theory*, edited by Ken Hirschkop and David Shepherd, 177–200. Manchester: Manchester University Press, 2001.

Derrida, Jacques. *Of Grammatology*. Translated by Gayatri Chakravorty Spivak. Baltimore: Johns Hopkins University Press, 2016.

Dhillon, Kiran Kaur, and Surita Mogan. "Language-Based Approaches to Understanding Literature: A Creative Activity Module." *English Teacher* 43, no. 2 (2014): 63–78.

Dimitri, Carl J. "*Go Down, Moses* and *Intruder in the Dust*: From Negative to Positive Liberty." *Faulkner Journal* 19, no. 1 (2003): 11–26.

Dixon, Thomas, Jr. *The Clansman: An Historical Romance of the Ku Klux Klan*. New York: Grosset and Dunlap, 1905.

———. *The Leopard's Spots: A Romance of the White Man's Burden, 1865–1900*. New York: Doubleday, Page and Company, 1902.

Dostoyevsky, Fyodor. *Crime and Punishment*. Translated by Oliver Ready. New York: Penguin Books, 2015.

Doyle, Don H. *Faulkner's County: The Historical Roots of Yoknapatawpha*. Chapel Hill: University of North Carolina Press, 2001.

Drainville, André C. *Contesting Globalization: Space and Place in the World Economy*. London: Routledge, 2004.

Duclos, Donald Philip. *Son of Sorrow: The Life, Works and Influence of Colonel William C. Falkner, 1825–1889*. San Francisco: International Scholars Publications, 1998.

Dussere, Erik. "The Debts of History: Southern Honor, Affirmative Action, and Faulkner's *Intruder in the Dust*." *Faulkner Journal* 17, no. 1 (2001): 37–57.

Egerton, John. *The Americanization of Dixie: The Southernization of America*. New York: Harper's Magazine Press, 1974.

Ellison, Ralph. "Change the Joke and Slip the Yoke." 1958. In *The Collected Essays of Ralph Ellison*, edited by John Callahan, 100–112. New York: The Modern Library, 1995.

Erdrich, Louise. *The Round House*. New York: HarperCollins, 2012.

Falkner, William C. *The Little Brick Church*. 1882. Upper Saddle River, NJ: Literature House, 1970.

———. *Rapid Ramblings in Europe*. Philadelphia: J. B. Lippincott, 1884. https://archive.org/details/rapidramblingsin00falk.

———. *The Siege of Monterey: A Poem*. Cincinnati: By the Author, 1851. https://archive.org/details/siegeofmonterey00falk.

———. *The Spanish Heroine: A Tale of War and Love*. Cincinnati: I. Hart, 1851. https://books.google.com.au/books/about/The_Spanish_Heroine_a_Tale_of_War_and_Lo.html?id=TpUIAQAAIAAJ.

———. *The White Rose of Memphis*. 1881. Reprinted with introduction by Robert Cantwell. New York: Coley Taylor, 1953.

Faulkner, William. *Absalom, Absalom!* 1936. New York: Vintage International, 1990.

———. *As I Lay Dying*. 1930. In *William Faulkner: Novels 1926–1929*, 1–178. New York: Library of America, 2006.

———. *Essays, Speeches and Public Letters*. 1965. New York: Modern Library, 2004.

———. *A Fable*. 1954. In *William Faulkner: Novels 1930–1935*, 665–1072. New York: Library of America, 1985.

———. *Faulkner at Nagano*. Edited by Robert A. Jelliffe. Tokyo: Kenkyusha, 1956.

———. *Faulkner in the University* 1959. Edited by Frederick L. Gwynn and Joseph L. Blotner. Charlottesville: University Press of Virginia, 1995.

———. *Flags in the Dust*. Originally published as *Sartoris* in 1929. In *William Faulkner: Novels 1926–1929*, 541–876. New York: Library of America, 2006.

———. *Go Down, Moses*. 1942. New York: Vintage International, 1990.

———. *The Hamlet*. 1940. In *William Faulkner: Novels 1936–1940*, 727–1075. New York: Library of America, 1990.

———. *Intruder in the Dust*. 1948. New York: Vintage International, 1991.

———. "Letter to a Northern Editor." 1956. In *Essays, Speeches and Public Letters*, 86–91.

———. "A Letter to the Leaders in the Negro Race." 1956. In *Essays, Speeches and Public Letters*, 107–112.

———. *Light in August*. 1932. In *William Faulkner: Novels 1930–1935*, 399–774. New York: Library of America, 1985.

———. *Lion in the Garden: Interviews with William Faulkner 1926–1962*. 1968. Edited by James B. Meriwether and Michel Millgate. Lincoln: University of Nebraska Press, 1980.

———. *The Mansion*. 1959. In *William Faulkner: Novels 1957–1962*, 327–722. New York: Library of America, 1999.

———. *The Marble Faun* and *A Green Bough*. 1924, 1933. New York: Random House, 1960.

———. *Mosquitoes*. 1927. In *William Faulkner: Novels 1926–1929*, 257–540. New York: Library of America, 2006.

———. "On Privacy." 1955. In Faulkner, *Essays, Speeches and Public Letters*, 62–75.

———. *The Reivers*. 1962. In *William Faulkner: Novels 1957–1962*, 723–971. New York: Library of America, 1999.

———. *Sanctuary*. 1931. New York: Vintage International, 1993.

———. *Selected Letters of William Faulkner*. Edited by Joseph Blotner. New York: Random, 1977.

———. *The Sound and the Fury*. 1929. In *William Faulkner: Novels 1926–1929*, 877–1141. New York: Library of America, 2006.

———. *The Sound and the Fury: A Hypertext Edition*. Edited by R. P. Stoicheff, Allison Muri, Joel Deshaye, Maria Truchan-Tataryn, Jon Bath, Dave Mitchell, and Kathleen Murphy. Updated February 2011. University of Saskatchewan. http://drc.usask.ca/projects/faulkner/.

———. *The Town*. 1959. In *William Faulkner: Novels 1957–1962*, 1–326. New York: Library of America, 1999.

———. "Wash." 1934. In *Collected Stories of William Faulkner*, 535–50. New York: Vintage Books, 1977.

Flowers, R. Barri. *Domestic Crimes, Family Violence and Child Abuse: A Study of Contemporary American Society*. Jefferson: McFarland, 2000.

Foucault, Michel. *The Archaeology of Knowledge and the Discourse on Language*. Translated by A. M. Sheridan Smith. New York: Vintage Books, 2010.

———. "What Is an Author?" In *The Foucault Reader*, edited by Paul Rabinow, translated by Josué V. Harari, 101–20. New York: Pantheon, 1984.

Freud, Sigmund. *Moses and Monotheism.* 1939. In *The Standard Edition of the Complete Psychological Works of Sigmund Freud*, translated by James Strachey, 23: 1–138. London: Vintage, 2001.

Frank, Joseph. *The Idea of Spatial Form.* New Brunswick: Rutgers University Press, 1991.

Friedman, Thomas L. *The World Is Flat: A Brief History of the Twenty-first Century.* 2005. New York: Picador, 2007.

Gajdusek, Linda. "Toward Wider Use of Literature in ESL: Why and How." *TESOL Quarterly* 22, no. 2 (1988): 227–57.

Gergen, Kenneth J. *The Saturated Self: Dilemmas of Identity in Contemporary Life.* 1991. New York: Basic Books, 2000.

Gerstenberger, Donna. "Meaning and Form in *Intruder in the Dust.*" *College English* 23, no. 3 (1961): 223–25.

Giles, Paul. *The Global Remapping of American Literature.* Princeton: Princeton University Press, 2011.

Gilster, Paul. *Digital Literacy.* New York: Wiley, 1997.

Glasgow, Ellen. *Barren Ground.* 1925. San Diego: Hartcourt, 1985.

Glicksberg, Charles I. Review of *Intruder in the Dust*, by William Faulkner. *Arizona Quarterly* 5 (1949): 85–88.

Godden, Richard. *Fictions of Labor: William Faulkner and the South's Long Revolution.* Cambridge: Cambridge University Press, 1997.

———. *William Faulkner: An Economy of Complex Words.* Princeton: Princeton University Press, 2007.

Graham, Allison. *Framing the South: Hollywood, Television, and Race During the Civil Rights Struggle.* Baltimore: Johns Hopkins University Press, 2001.

Gray, Richard. "Foreword: Inventing Communities, Imagining Places: Some Thoughts on Southern Self-Fashioning." In *South to a New Place: Region, Literature, Culture*, edited by Suzanne W. Jones and Sharon Monteith, xiii-xxiii. Baton Rouge: Louisiana State University Press, 2002.

———. *Southern Aberrations: Writers of the American South and the Problems of Regionalism.* Baton Rouge: Louisiana State University Press, 2000.

Greenberg, Kenneth S. *Honor and Slavery: Lies, Duels, Noses, Masks, Dressing as a Woman, Gifts, Strangers, Humanitarianism, Death, Slave Rebellions, the Proslavery Argument, Baseball, Hunting, and Gambling in the Old South.* Princeton: Princeton University Press, 1996.

Guarneri, Carl. *America in the World: United States History in Global Context.* Boston: McGraw-Hill, 2007.

Habermas, Jürgen. *The Theory of Communicative Action.* Vol. 1, *Reason and the Rationalization of Society.* Translated by Thomas McCarthy. Boston: Beacon Press, 1984.

Hagood, Taylor. *Faulkner's Imperialism: Space, Place, and the Materiality of Myth.* Baton Rouge: Louisiana State University Press, 2008.

Harris, Joel Chandler. *Uncle Remus: His Songs and His Sayings.* 1880. New York: Penguin, 1986.

Haucap, Justus and Ulrich Heimeshoff. "Google, Facebook, Amazon, eBay: Is the Internet Driving Competition or Market Monopolization?" *International Economics & Economic Policy* 11, nos. 1–2 (2014): 49–61. doi:10.1007/s10368–013–0247–6.

Hegel, G. W. F. *Phenomenology of Spirit*. Translated by A. V. Miller. Oxford: Oxford University Press, 1977.

Hemingway, Earnest. *The Sun Also Rises*. 1926. New York: Scribner, 2006.

Henry, O. "A Municipal Report." 1909. In *Selected Stories from O. Henry*, edited by C. Alphonso Smith, 224–41. New York: Odyssey, 1922.

Hentz, Caroline Lee. *The Planter's Northern Bride*. Philadelphia: T. B. Peterson and Brothers, 1854. https://archive.org/details/plantersnorthern00hent.

Hermans, Hubert J.M. and Harry J. G. Kempen. *The Dialogical Self: Meaning as Movement*. San Diego: Academic Press, 1993.

Herriman, Michael. "Autonomous Learning in EFL in Japan." *International Journal of Learning* 14, no. 5 (2007): 141–47.

Hickerson, Thomas Felix. *The Falkner Feuds*. Chapel Hill, NC: The Colonial Press, 1964.

Hicks, Granville. Review of *Intruder in the Dust*, by William Faulkner. *Tomorrow* 8 (1949): 57–58.

Hirschkop, Ken. *Mikhail Bakhtin: An Aesthetic for Democracy*. Oxford: Oxford University Press, 1999.

Hirschkop, Ken, and David Shepherd, eds. *Bakhtin and Cultural Theory*. 1989. Manchester: Manchester University Press, 2001.

Hoar, Jere. "Book Reading in the Senior Years: The Habits and Preferences of 200 Mississippians." *Journal of Educational Sociology* 34, no. 3 (1960): 137–44.

Hobson, Fred, ed. *South to the Future: An American Region in the Twenty-First Century*. Athens: University of Georgia Press, 2002.

Hodes, Martha. *White Women, Black Men: Illicit Sex in the Nineteenth-Century South*. New Haven: Yale University Press, 1997.

Hodge, Matthew. "'Bicycles, Muscles, Cigarettes,' and ESL: Raymond Carver in the Classroom Matthew Hodge, Central Piedmont Community College." *Teaching American Literature* 7, nos. 1–2 (2014): 44–52.

Hoffman, Frederick J. *William Faulkner*. 1961. New Haven: College and University Press, 1966.

———. "William Faulkner: An Introduction." In Hoffman, *William Faulkner: Three Decades*, 1–50.

Hoffman, Frederick J., and Olga W. Vickery, eds. *William Faulkner: Three Decades of Criticism*. 1960. New York: Harbinger, 1963.

Hooks, bell. *Teaching to Transgress: Education as the Practice of Freedom*. New York: Routledge Press. 1994.

Howard, Tharon W. *Design to Thrive: Creating Social Networks and Online Communities that Last*. Amsterdam: Elsevier, 2010.

Howe, Irving. *William Faulkner: A Critical Study*. 1952. 4th ed. Chicago: Elephant Paperbacks, 1991.

Hulle, Dirk Van. "Hypertext and *Avant-texte* in Twentieth-Century and Contemporary Literature." In *A Companion to Digital Literary Studies*, edited by Ray Siemens and Susan Schreibman, 139–59. Malden, MA: Blackwell, 2007.

Hurston, Zora Neale. *Their Eyes Were Watching God.* 1937. New York: Harper Perennial, 2006.

Intruder in the Dust. Directed by Clarence Brown. 1949. Burbank, CA: Warner Home Video, 2011. Videodisc (DVD), 87 min.

Irwin, John T. *Doubling and Incest/Repetition and Revenge: A Speculative Reading of Faulkner.* 1975. Baltimore: Johns Hopkins University Press, 1996.

Jehlen, Myra. *Class and Character in Faulkner's South.* New York: Columbia University Press, 1976.

Jones, Gavin. *American Hungers: The Problem of Poverty in U.S. Literature, 1840–1945.* Princeton: Princeton University Press, 2008.

Jones, Suzanne W., and Sharon Monteith. *South to a New Place: Religion, Literature, Culture.* Baton Rouge: Louisiana State University Press, 2002.

Kanetsuna, Tomoyuki, and Peter K. Smith. "Pupil Insights into Bullying, and Coping with Bullying: A Bi-National Study in Japan and England." *Journal of School Violence* 1, no. 3 (2002): 5–29.

Karaganis, Joe, "Negotiating the National Voice in Faulkner's Late Work." In *A Gathering of Evidence: Essays on William Faulkner's "Intruder in the Dust,"* edited by Michel Gresset and Patrick Samway, S.J., 97–129. Philadelphia: St. Joseph's University Press, 2004.

King, Martin Luther, Jr. "'I Have a Dream': Speech Delivered at the March on Washington, August 28, 1963." In Sundquist, *King's Dream,* 229–34.

King, Richard H. *A Southern Renaissance: The Cultural Awakening of the American South, 1930–1955.* Oxford: Oxford University Press, 1980.

Kreyling, Michael. *Inventing Southern Literature.* Jackson: University Press of Mississippi, 1998.

Kvale, Steiner. "Introduction: From the Archaeology of the Psyche to the Architecture of Cultural Landscapes." In Kvale, *Psychology and Postmodernism,* 1–16.

———, ed. *Psychology and Postmodernism.* London: Sage Publications, 1992.

Labbé, Jessica. "'Death by Misadventure': Teaching Transgression in/through Larsen's *Passing*" *College Literature* 37, no. 4 (2010): 120–44.

Ladd, Barbara. *Nationalism and the Color Line in George W. Cable, Mark Twain, and William Faulkner.* Baton Rouge: Louisiana State University Press, 1996.

Landow, George P. *Hypertext 3.0: Critical Theory and New Media in an Era of Globalization.* Baltimore: Johns Hopkins University Press, 2006.

Lather, Patti. "Postmodernism and the Human Sciences." In Kvale, *Psychology and Postmodernism,* 88–109.

Leary, Lewis. *William Faulkner of Yoknapatawpha County.* New York: Thomas Y. Crowell, 1973.

Lee, Icy. "Working Smarter, Not Working Harder: Revisiting Teacher Feedback in the L2 Writing Classroom." *Canadian Modern Language Review* 67, no. 3 (2011): 377–99.

Leerssen, Joep. "Literary Historicism: Romanticism, Philologists, and the Presence of the Past." *Modern Language Quarterly* 65, no. 2 (2004): 221–43.

Li, Stephanie. "*Intruder in the Dust* from Novel to Movie: The Development of Chick Mallison." *Faulkner Journal* 16, no. 1–2 (2000–2001): 105–18.

Limón, José E. *American Encounters: Greater Mexico, the United States, and the Erotics of Culture*. Boston: Beacon, 1998.
López, Ian Haney. *White By Law: The Legal Construction of Race*. New York: New York University Press, 2006.
Mahmoud, Abdulmoneim. "Strategy-Based Peer Assistance in EFL Writing: An Alternative to Traditional Peer Correction." *Theory and Practice in Language Studies* 2, no.10 (2012): 1987–92.
Martin, Valerie. *Property*. New York: Vintage Books, 2007.
Matthews, John T. "The Southern Renaissance and the Faulknerian South." In Monteith, *The Cambridge Companion*, 116–31.
———. "This Race Which Is Not One: The 'More Inextricable Compositeness' of William Faulkner's South." In Smith and Cohn, *Look Away!* 201–26.
———. *William Faulkner: Seeing Through the South*. Chichester: Wiley-Blackwell, 2009.
McPherson, Tara. *Reconstructing Dixie: Race, Gender, and Nostalgia in the Imagined South*. Durham: Duke University Press, 2003.
Mead, G. H. *Mind, Self, and Society*. Chicago: University of Chicago Press, 1934.
Mencken, H. L. "The Sahara of the Bozart." 1917. In Mencken, *Prejudices: Second Series*, 136–54. New York: Alfred A. Knopf, 1920.
Metalious, Grace. *Peyton Place*. 1956. Boston: Northeastern University Press, 1999.
Miller, J. Hillis. *On Literature*. London: Routledge, 2002.
Millgate, Michael. *The Achievement of William Faulkner*. London: Constable, 1965.
Millichap, Joseph R. *Dixie Limited: Railroads, Culture, and the Southern Renaissance*. Lexington: University Press of Kentucky, 2002.
Miyazoe, Terumi, Terry Anderson, and Shinichi Sato. "'To-do-or-not-to-do dilemma' Online: Blog and Essay Writing Visualizations in a Blended-Format University English Course." *E-FLT: Electronic Journal of Foreign Language Teaching* 11 (2014): 165–82. http://e-flt.nus.edu.sg/v11s12014/miyazoe.pdf.
Monteith, Sharon, ed. *The Cambridge Companion to the Literature of the American South*. Cambridge: Cambridge University Press, 2013.
———. "Introduction: Mapping the Figurative South," in Monteith, *The Cambridge Companion*, 1–10.
Moreland, Richard C. "Contextualizing Faulkner's *Intruder in the Dust*: Sherlock Holmes, Chick Mallison, Decolonization, and Change." *Faulkner Journal* 12, no. 2 (1997): 57–68.
———. *Faulkner and Modernism: Reading and Rewriting*. Madison: University of Wisconsin Press, 1990.
Morris, Wesley, and Barbara Alverson Morris. *Reading Faulkner*. Madison: University of Wisconsin Press, 1990.
Morrison, Toni. *Beloved*. 1987. New York: Vintage International, 2004.
———. *Home*. London: Vintage Books, 2013.
Mortimer, Gail L. *Faulkner's Rhetoric of Loss: A Study in Perception and Meaning*. Austin: University of Texas Press, 1983.
Muir, John. *My First Summer in the Sierra*. 1911. New York: Penguin, 1987.
Murphree, Alex. "Faulkner's Novel Offers Parable on Southerner's Duty to Negroes." Review of *Intruder in the Dust*, by William Faulkner. *Post* (October 3, 1948): C4.

Nakamura, Akira, and Masao Kikuchi. "What We Know, and What We Have Not Yet Learned: Triple Disasters and the Fukushima Nuclear Fiasco in Japan." *Public Administration Review* 71, no. 6 (2011): 893–99. doi:10.1111/j.1540–6210.2011.02437.x.

Nambu, Saori, Ayako Nasu, Shigeru Nishimura, Akiyoshi Nishimura, and Satoshi Fujiwara. "Fatal Child Abuse in Japan: Does a Trend Exist toward Tougher Sentencing?" *Journal of Injury and Violence Research* 3, no. 2 (2011): 74–79. doi:10.5249/jivr.v3i2.73.

O'Brien, Michael. *The Idea of the American South, 1920–1941*. Baltimore: Johns Hopkins University Press, 1979.

O'Connor, William Van. *The Tangled Fire of William Faulkner*. 1954. New York: Gordian Press, 1968.

Oster, Judith. "Seeing with Different Eyes: Another View of Literature in the ESL Class" *TESOL Quarterly* 23, no. 1 (1989): 85–103.

Padgett, John B. *William Faulkner on the Web*. University of Mississippi, 1995. http://www.mcsr.olemiss.edu/%7Eegjbp/faulkner/faulkner.html.

Paech, Venessa. "The Inevitable Exile: A Missing Link in Online Community Discourse" In Brabazon, *Digital Dialogues*, 11–41.

Page, Thomas Nelson. "Marse Chan: A Tale of Old Virginia" In *In Ole Virginia; or, Marse Chan and Other Stories*, 1–38. 1887. Nashville: J.S. Sanders, 1991.

Petrone, Robert and Robert Gibney. "The Power to Speak and Listen: Democratic Pedagogies for American Literature Classrooms," *English Journal* 94, no. 5 (2005): 35–39.

Poe, Edgar A. "The Purloined Letter." In *Tales*, 200–18. New York: Wiley and Putnam, 1845.

Polk, Noel. *Children of the Dark House: Text and Context in Faulkner*. Jackson: University Press of Mississippi, 1996.

Puxan, Marta. "Narrative Strategies on the Color Line: The Unreliable Narrator Shreve and Racial Ambiguity in Faulkner's *Absalom, Absalom!*" *Mississippi Quarterly* 60, no. 3 (2007): 529–59.

Rai, Asha. "Use of Literature in Teaching English." *International Journal of Educational Research and Technology* 3, no. 3 (2012): 71–80. doi.10.1016/j.sbspro.2012.05.159.

Railton, Stephen. *Digital Yoknapatawpha*. University of Virginia. http://faulkner.iath.virginia.edu/.

———. *Faulkner at Virginia: An Audio Archive*. University of Virginia. http://faulkner.lib.virginia.edu/.

Samway, Patrick H., S.J. *Faulkner's "Intruder in the Dust": A Critical Study of the Typescripts*. Troy, NY: Whitston, 1980.

Schmidt, Peter. *Sitting in Darkness. New South Fiction, Education, and the Rise of Jim Crow Colonialism, 1865–1920*. Jackson: University Press of Mississippi, 2008.

Schwartz, Lawrence H. *Creating Faulkner's Reputation: The Politics of Modern Literary Criticism*. 1988. Knoxville: University of Tennessee Press, 1990.

Scott, Walter. *Ivanhoe*. 1820. New York: American Book Company, 1904. https://archive.org/details/ivanhoe01scotgoog.

Sen, Amartya. *Identity and Violence: The Illusion of Destiny*. New York: W. W. Norton, 2006.

Seng, Goh Hock, and Fatimah Hashim. "Use of L1 in L2 Reading Comprehension among Tertiary ESL Learners." *Reading in a Foreign Language* 18, no. 1 (2006): 29–54. http://nflrc.hawaii.edu/rfl/April2006/abstracts.html#goh.

Sharp, Stephen. *A New Look at the Interactive Writing Classroom: Methods, Strategies, and Activities to Engage Students*. Lanham: Rowan and Littlefield Education, 2011.

Smith, Jon. *Finding Purple America: The South and the Future of American Cultural Studies*. Athens: University of Georgia Press, 2013.

Smith, Jon, and Deborah Cohn. "Introduction: Uncanny Hybridities." In Smith and Cohn, *Look Away!*, 1–19.

———, eds. *Look Away!: The U.S. South in New World Studies*. Durham: Duke University Press, 2004.

Solomon, Gwen, and Lynne Schrum. *Web 2.0: New Tools, New Schools*. Eugene: International Society for Technology in Education, 2007.

Southworth, E.D.E.N. *The Hidden Hand*. 1888. New York: Burt, n.d. https://archive.org/details/hiddenhand00soutuoft.

Stoddard, Elizabeth. *The Morgesons*. 1862. New York: Penguin, 1984.

———. *Two Men*. 1865. Lincoln, NE: University of Nebraska Press, 2008.

Stowe, Harriet Beecher. *Uncle Tom's Cabin; or, Life among the Lowly*. 1852. New York: W.W. Norton, 1994.

Sugimori, Masami. "Signifying, Ordering, and Containing the Chaos: Whiteness, Ideology, and Language in Intruder in the Dust." *Faulkner Journal* 22, no. 1–2 (2006/2007): 54–73.

Sundquist, Eric J. *King's Dream*. New Haven: Yale University Press, 2009.

———. *Faulkner: The House Divided*. Baltimore: Johns Hopkins University Press, 1983.

Suppasetseree, Suksan and Nutprapha K. Dennis. "The Use of Moodle for Teaching and Learning English at Tertiary Level in Thailand." *International Journal of the Humanities* 8, no. 6 (2010): 29–46.

Suzuki, Tatsujiro. "Deconstructing the Zero-Risk Mindset: The Lessons and Future Responsibilities for a Post-Fukushima Nuclear Japan." *Bulletin of the Atomic Scientists* 67, no. 5 (2011): 9–18. doi:10.1177/0096340211421477.

Tate, Allen. "Remarks on the Southern Religion." In *I'll Take My Stand*, by Twelve Southerners, 155–75. 1930. Baton Rouge: Louisiana State University Press, 2006.

Tettegah, Sharon. "Computer Mediated Social Justice: A New Model for Educators." *TechTrends* 46, no. 5 (2002): 27–33.

The Birth of a Nation. Directed by D. W. Griffith. 1915. New York: Kino Classics, 2011. Blue-ray, 187 min.

Thoreau, Henry David. *Walden; or, Life in the Woods*. Boston: Ticknor and Fields, 1854. https://archive.org/details/waldenorlifeinwo1854thor.

Tikiz, Gülşah, and Feryal Çubukçu. "Symbiotic Interface between Literature, Culture and Language Learning." *Journal of International Social Research* 6, no. 26 (2013): 134–43.

Tocqueville, Alexis de. *Democracy in America*. 1835, 1840. Translated by Arthur Goldhammer. New York: Library of America, 2004.

Todorov, Tzvetan. *Mikhail Bakhtin: The Dialogical Principle*. 1981. Translated by Wlad Godzich. Minneapolis: University of Minnesota Press, 1984.

Traore, Moussa, and Lydia Kyei-Blankson. "Using Literature and Multiple Technologies in ESL Instruction." *Journal of Language Teaching and Research* 2, no. 3 (2011): 561–68. http://www.editlib.org/p/35485.

Tuan, Yi-Fu. "Introduction: Cosmos versus Hearth." In *Textures of Place: Exploring Humanist Geographies*, edited by Paul C. Adams, Steven Hoelscher, and Karen E. Till, 319–25. Minneapolis: University of Minnesota Press, 2001.

Twain, Mark. *Adventures of Huckleberry Finn*. 1885. In *Mark Twain: Mississippi Writings*, edited by Guy Carwell, 617–912. New York: Library of America, 1982.

Twelve Southerners. *I'll Take My Stand*. 1930. Baton Rouge: Louisiana State University Press, 2006.

Upton, Thomas A. "First and Second language Use in Reading Comprehension Strategies of Japanese ESL Students." *TESL-EJ* 3, no. 1 (1997): 1–23.

Vickery, Olga W. *The Novels of William Faulkner: A Critical Interpretation*. 1959. Revised Edition. Baton Rouge: Louisiana State University Press, 1964.

Waddle, Mary Betsy. "A Late Dinner with the Faulkners." In *William Faulkner of Oxford*, edited by James W. Webb and A. Wigfall Green, 139–42. Louisiana: Louisiana State University Press, 1965.

Warren, Robert Penn. *All the King's Men*. 1946. San Diego: Harcourt, 2002.

Weingourt, Rita, T. Maruyama, T., I. Sawada, and J. Yoshino. "Domestic Violence and Women's Mental Health in Japan." *International Nursing Review* 48, no. 2 (2001): 102–8. doi:10.1046/j.1466–7657.2001.00059.x.

Weinstein, Philip M. *Faulkner's Subject: A Cosmos No One Owns*. Cambridge: Cambridge University Press, 1992.

———. *What Else But Love? The Ordeal of Race in Faulkner and Morrison*. New York: Columbia University Press, 1996.

Weiss, Robert J. *"We Want Jobs": A History of Affirmative Action*. New York: Garland, 1997.

Wells, Jeremy. *Romances of the White Man's Burden: Race, Empire, and the Plantation in American Literature, 1880–1936*. Nashville: Vanderbilt University Press, 2011.

Williamson, Joel. *William Faulkner and Southern History*. New York: Oxford University Press, 1993.

Winchell, Mark Royden. *Cleanth Brooks and the Rise of Modern Criticism*. Charlottesville: University Press of Virginia, 1996.

———. *Reinventing the South: Versions of a Literary Region*. Columbia: University of Missouri Press, 2006.

Wolfe, Paula. "'The Owl Cried': Reading Abstract Literary Concepts with Adolescent ESL Students." *Journal of Adolescent and Adult Literacy* 47, no. 5 (2004): 402–13.

Yaeger, Patricia. *Dirt and Desire: Reconstructing Southern Women's Writing, 1930–1990*. Chicago: University of Chicago Press, 2000.

Yaman, Saziye. "Technology Supported Learning Platform: Moodle Integrated Academic Course." *Turkish Online Journal of Distance Education* 11, no. 2 (2010): 146–60.

Young, Stark. "Golden Apples." *English Journal* 26, no. 9 (1937): 689–97.

Zender, Karl F. *The Crossing of The Ways: William Faulkner, the South, and the Modern World.* New Brunswick: Rutgers University Press, 1989.

Index

About the Author

Koichi Fujino is professor of American literature and culture at Seinan Gakuin University, Japan, where he teaches and researches American literature and culture, the literature of the American South, American Modernism, and English as a second language.

Lightning Source UK Ltd.
Milton Keynes UK
UKHW040350181218
334007UK00026B/266/P

9 781498 547475